Other books by Theoni Pappas

THE JOY OF MATHEMATICS
MORE JOY OF MATHEMATICS
MATH-A-DAY
MATHEMATICAL FOOTPRINTS
THE MATHEMATICS CALENDAR
MAGIC OF MATHEMATICS
MATHEMATICAL SCANDALS
THE MUSIC OF REASON
MATHEMATICS APPRECIATION
MATH TALK
WHAT DO YOU SEE? An Optical Illusion Slide Show
GREEK COOKING FOR EVERYONE
LET'S DANCE THE GREEK WAY

Other children's books by Theoni Pappas

FRACTALS, GOOGOLS & OTHER MATHEMATICAL TALES
THE CHILDREN'S MATHEMATICS CALENDAR
THE ADVENTURES OF PENROSE—THE MATHEMATICAL CAT

Math for Kids
& Other People Too!

by theoni pappas

— Wide World Publishing/Tetra —

Portions of this book have appeared in *The Children's Mathematics Calendar.*

Wide World Publishing/Tetra
P.O. Box 476
San Carlos, CA 94070

websites:
http://www. wideworldpublishing.com
http://www. mathproductsplus.com

Printed in the United States of America.

2nd Printing April 2001

Library of Congress Cataloging-in-Publication Data

Pappas, Theoni
Math for kids & other people too! / Theoni Pappas. __ 1st ed.
p. cm.
Summary : Explores mathematics through stories, puzzles, challenges, games, tricks, and experiments. Answers provided in a separate section.
ISBN 1-84550-13-4 (alk. paper)
1. Mathematics--Study and teaching (Elementary) [1. Mathematical recreations.] I. Title
QA135.5.P3325 1997
793.7'4--dc21

97-43091
CIP
AC

For two very special people in my life

Eli and my sister Pearl

Preface

Don't be afraid to open this book at random. It is designed to be used that way. Each chapter is independent of the others. Whenever you feel like reading a math story open up to something in the first part of the book. If you would like to tackle a puzzle, play a game or learn a math trick, look in the second part of the book.

Don't hesitate tackling the questions, problems, experiments, and researching part of the chapter. They are designed to help you discover ideas and have fun. If you come across a question or problem you can't answer, try not to look at the answer section until you have let that question churn in your mind for a few days. The solution may just need time for you to think it through.

—Theoni Pappas

TABLE OF CONTENTS

MATH STORIES & IDEAS

TABLE OF CONTENTS continued

MATH PUZZLES GAMES & TRICKS

MATH STORIES and IDEAS

How the fractions squeezed between the counting numbers.

Fractions are a strange breed of numbers. When they were first invented,

"How dare you put 1 on top of 2," 2 exclaimed. "Why didn't you put 2 on top of 1? like this, 2/1?" 2 asked.

the counting numbers looked down their noses at them. 1, 2, 3, ... and all the other counting numbers were appalled when the first fraction "1/2" appeared.

"How dare you put 1 on top of 2!" 2 exclaimed. "Why didn't you put 2 on top of 1? like this, 2/1?" 2 asked.

"Then the number formed would not be a fraction of 1, " 1/2 declared. "It would have just been another way of writing you, 2," 1/2 explained.

After 1/2 appeared, there was no holding back the other fractions whose values were less than 1, such as

1/2, 1/3, 1/4, ...

2/3, 2/4, 2/5, ...

3/4, 3/5, 3/6, ...

On and on they marched in front of the counting numbers. Some, such as 1/2 and 2/4, named the same

$$\frac{1}{2}$$

"It would have just been another way of writing you, 2."

amount, but that didn't prevent them from appearing.

It seemed that once the fractions were defined by mathematicians, they became indispensable. Mathematicians often wondered how they had been able to work without them. Fractions became the fad.

One can understand how hurt the counting numbers felt. Once they had

"You are an improper fraction!"

{1, 2, 3, 4, 5, 6, 7, 8, 9, 10, 11, 12, 13, 14 . . .}

been the only numbers around – they were top dog – now they had to share their space. They were no longer the only numbers that could solve problems. Little did the counting numbers know that the future held many new types of numbers which would crowd them even more.

Even though the counting numbers had to accept the fractions, they refused to mix with them. They never invited the fractions to any of their counting parties. The fractions were not allowed to participate in any of the counting numbers count offs because

"Call me what you want, I am still as useful as you whole numbers."

it is impossible to list two consecutive fractions because one can always find more fractions between any two fractions.

$$\frac{3}{2}$$

The counting numbers felt they were neater, more organized, more refined than the fractions, especially because between any two consecutive counting numbers no others existed.

At about the time the counting numbers and fractions were at least tolerating one another, a fraction greater than 1 appeared. Apparently a mathematician had posed the problem– if I have 1 and 1/2 apple pies and each whole pie is to be cut into halves, how many halves will 1 1/2 pies give? –3 halves, which was written 3/2.

When this happened, you could hear all the counting numbers shout in unison–"The 3 on top of the 2! How improper! 3/2, you are an improper fraction," they decreed. But their name calling didn't deter other improper fractions from appearing. So 5/4, 6/5, 7/6, and 96/43, and any fraction in which the top was larger than the bottom was declared by the counting numbers as ***improper*** and top heavy–and they did not want to even look at them.

But the counting numbers were in for a big surprise because in 3/2, the 2 got tired of carrying the larger number 3 and said, "Since we were made from 1 and a half pies, why can't we rewrite ourselves as 1 1/2 .

When the counting numbers saw this they shouted "Stop! Stop! Don't mix us up with fractions. We are different– distinct." But it was too late. A whole new group of fractions formed, and predictably they were called ***mixed numbers***. Since mixed numbers (such as 3 2/3, 10 16/51) combine a counting number and a fraction, the counting numbers and the fractions were forced to spend more time together. In fact, infinitely many became good friends. After all , they all had a common purpose, to be ready to solve problems!

puzzling questions

1. Between which two whole number does the fraction 23/4 lie?

a) 1 and 2
b) 2 and 3
c) 4 and 5
d) 5 and 6
e) 7 and 8
f) none of the above

2. Between which two whole number does the fraction 1/104 lie?

3. How many fractions are there between 1 and 2?

a) 4
b) 10
c) 100
d) 937
e) infinite amount

4. Which fraction in each pair is larger?

a) 1/2 or 2/3
b) 5/4 or 4/5
c) 1/100 or 1/10
d) 6/3 or 7/6

5. Arrange the numbers below in ascending order on the number line.

2/3 1/2 5/4 1/4 1/12 1 2/3 2 1/3 1 2/6

0 1 2

Dominoes discover Polyominoes

" Stop! Why are you breaking up the domino set into little squares?" Dominoes demanded.

"I am conducting an experiment of the highest order, " Mr. Far Out replied. " Why should we just have dominoes? Why not other *ominoes*? I believe dominoes are only one small part of a much greater world, the world of polyominoes. Are you willing to join me in this experiment?" he asked Dominoes?

"We guess we have no choice, since you have started, and we are already in your hands. Besides it would be good to find other ominoes to share our squares with," Dominoes replied. "So how will you go about this?" they asked Mr. Far Out.

" I just separate all your dominoes into the two squares that make up each piece. I turn all the squares over so no dots show because those dots are for another game. Now I decide, or should I say WE decide the rules for making a polyomino. Do you have any ideas, Dominoes?"

"It's simple!" Dominoes said. Mr. Far Out looked at Dominoes skeptically. "Start with just one square and call it an monomino. Two, of course, will be a domino, three a tromino, and so forth and that is all there is to it." Dominoes felt very accomplished with their proposition.

"Just one minute," said Mr. Far Out. "What about a shape like this? What would you call it?" he asked.

"Gee, we had not thought of that. Oh dear, you are right, Mr. Far Out. There are a lot more possibilities for four squares, five squares and on and on. What do we do?" asked Dominoes.

Mr. Far Out paused a moment, then

asked, "Why don't we say that a tromino

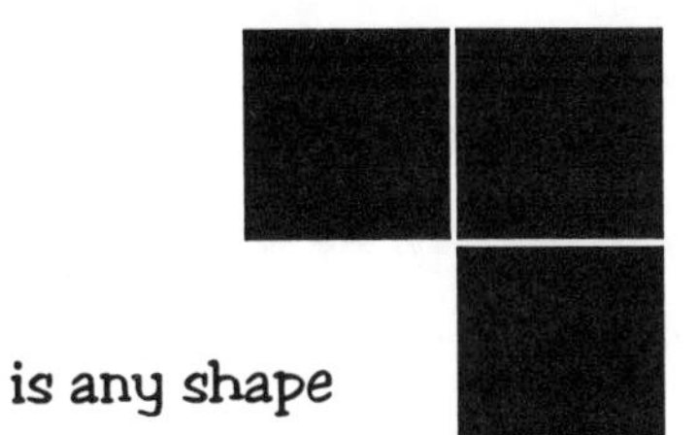

is any shape made from three squares joined at common edges. So these two would be the only tromino shapes.

This is not a tromino because an edge is shared by 2 squares.

This is not a tromino since 2 of the squares are only joined at a vertex point."

"That is a great idea. You know you're not that far out, Mr. Far Out," Dominoes said. "This means we are part of a family of polyominoes."

"In addition, it means there are new games and shapes to discover. Let's get acquainted with some of these," Mr. Far Out suggested.

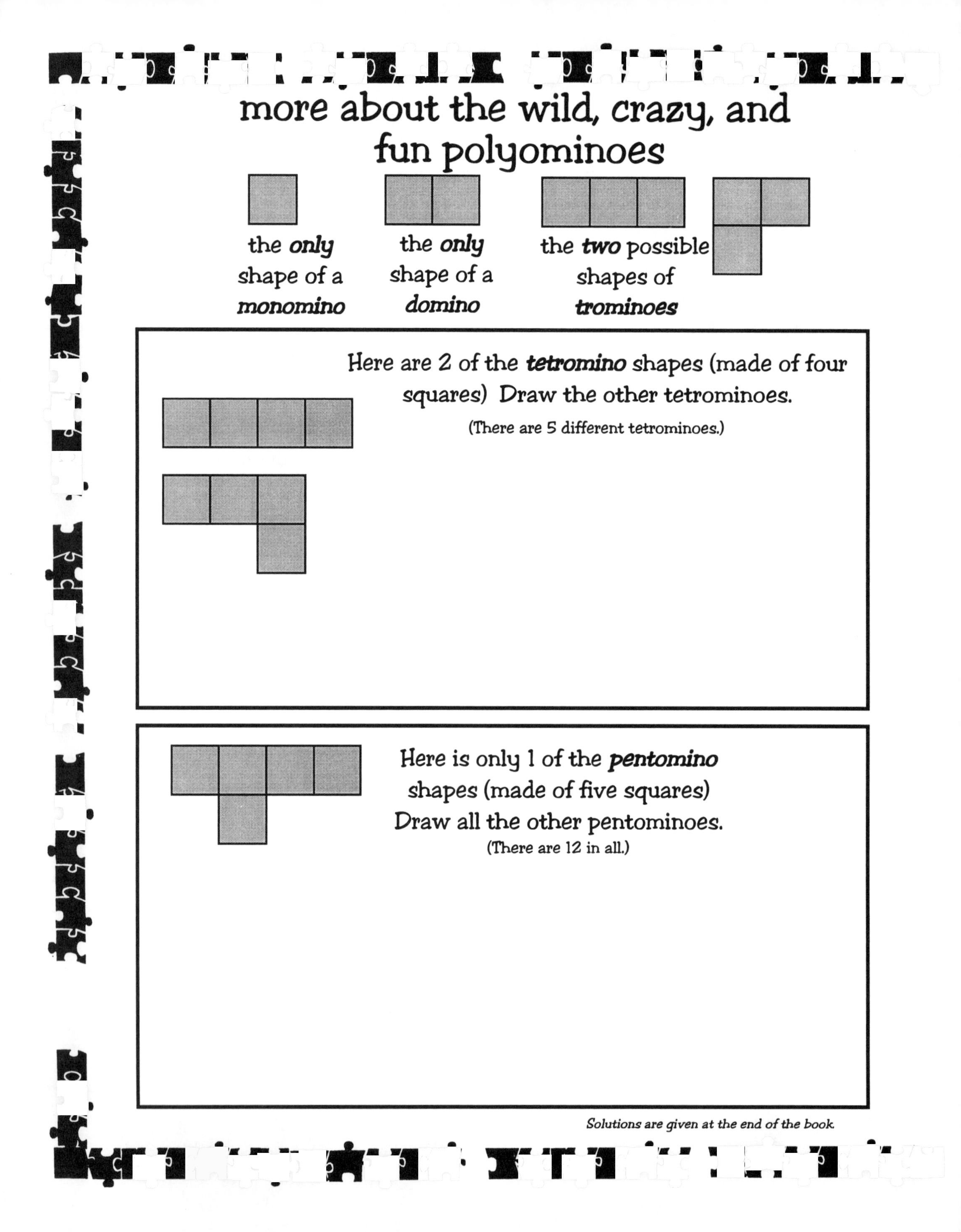

more about the wild, crazy, and fun polyominoes

the *only* shape of a ***monomino***

the *only* shape of a ***domino***

the ***two*** possible shapes of ***trominoes***

Here are 2 of the ***tetromino*** shapes (made of four squares) Draw the other tetrominoes.

(There are 5 different tetrominoes.)

Here is only 1 of the ***pentomino*** shapes (made of five squares) Draw all the other pentominoes.

(There are 12 in all.)

Solutions are given at the end of the book.

A polyomino game

Take a piece of graph paper and make any size rectangle. Choose one of the polyominoes and see if all the squares of the rectangle can be covered with the shapes of that polyomino.

* * *

See if you can discover a method for determining when a polyomino will cover a certain shape.

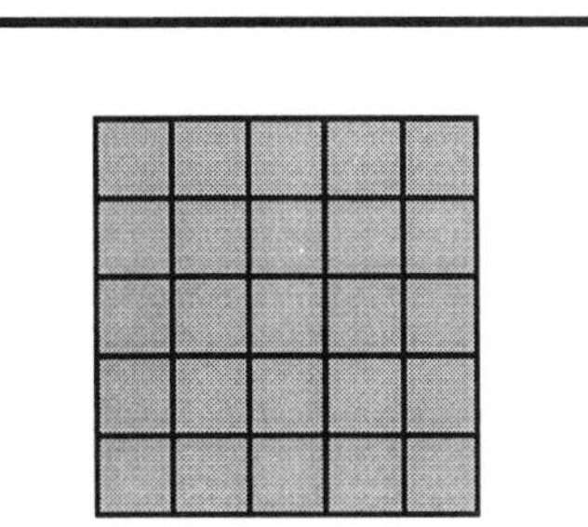

Which of these rectangles can be covered with tetromino shapes?

Remember to use only tetrominoes with no overlapping or empty squares.

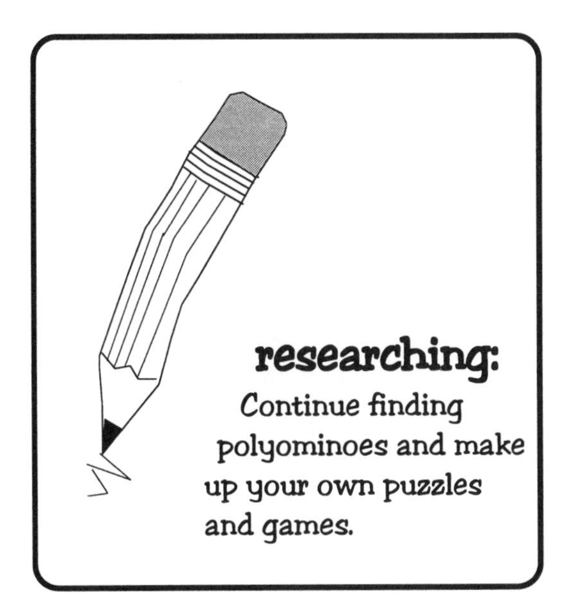

researching:

Continue finding polyominoes and make up your own puzzles and games.

Solutions are given at the end of the book.

Palindromes – the forward & backwards numbers

"Palindromes are sure a weird bunch of numbers," 25 said in a cutting voice.

"I agree totally," 297 said. "Look at those three palindromes over there," it added, pointing to 353, 1551 and 79497. "You never know whether they are coming or going, since they read the same from left to right or right to left." 297 laughed as it spoke.

Hearing them, zero came over. "Why are you always giving the palindromes a hard time?" zero asked 25 and 297. Zero felt sympathetic toward the palindromes, since zero itself had undergone much ridicule from the other numbers for hundreds of years. They used to tease zero, calling it a worthless number. "Granted 25, you are a perfect square number, but you're still an odd number. And you 297, even though you are a composite number you're still an odd number.

Stop bothering the palindromes," zero said emphatically.

But this did not discourage the badgering. "Listen 1551, do you know if you're coming or going," taunted 6, one of the perfect numbers.

You never know of whether they are coming or going.

This time 353 got angry. "None of you has the right to make fun of us, especially since we are all formed from numbers like you," 353 roared.

"Oh, prove it," 297 yelled back.

"No problem," 79497 said with a smile on its face.

Why are you always giving palindromes a hard time?

"Take 297. Reverse it digits, 792. Now add these two numbers together and you get 1089. Reverse its digits again 9801.

None of you has the right to make fun of us.

Now by adding these two numbers again and continuing this process, eventually we get the palindrome 79497."

0

25 and 6 gasped. "A palindrome is made from you 297. We are not hanging around with you any more," they sneered.

"Don't be so cocky," 79497 said. "Palindromes can be made from any of you. Sometimes you have to continue the process longer, but all of you end up a palindrome!" This time the palindromes were the ones laughing.

"See," zero said with a big smile on its face," We're all one happy family."

Here's how 297 is turned into a palindrome.

297	
+792	here's 297 reversed
1089	
+9801	here's 1089 reversed
10890	here's the sum of 1089 & 9801
+09801	here's 10980 reversed
20691	here's the sum of 10890 & 9801
+19602	here's 20691 reversed
40293	here's the sum of 20691 & 19602
+39204	here's 20691 reversed
79497	

palindrome experiments

1. Each number below has a palindrome that can be made from it. Try making the palindrome for each of the following numbers. Some take longer, but each should end up a palindrome.

2. Circle which of the following numbers are not palindromes.

33 1345431 83

494 456

3. Palindromes do not have to be numbers. English words such as *dad,* reads the same when read from the right or the left. Even sentences can be palindromes such as, *Was it a rat I saw?.*

Can you think of some palindrome words? How about a palindrome sentence?

Solutions are at the back of the book.

researching:

Do you know the meaning of the following – *odd number, prime number, perfect number* and *perfect square number.* If not ask, your teacher, a friend or the librarian, or look it up in a book. Why not find out what makes them different?

HANDS UP symmetry & the art problem

So many things all around us have symmetry.

Symmetry is a state of balance. Let's just look at two kinds of symmetry—*line* symmetry and *point symmetry.* An object, such as this triangle has *line symmetry,* since one can find a line that divides it exactly in half, so that each half is the mirror image of the other.

But let's check to see if this triangle has line symmetry?

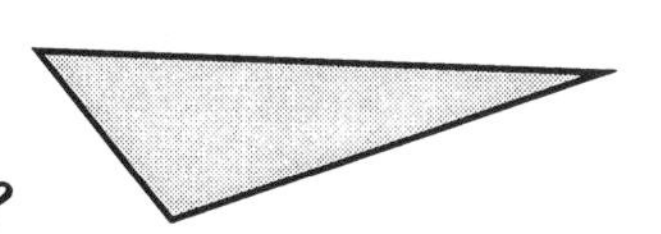

Can you find a way of dividing it with a line so that if you fold the triangle along that line both parts would match exactly?

No, it is not possible to find a line a symmetry for the second triangle.

Another way to test for line symmetry is to use a mirror and see, if by placing the mirror on the line of symmetry you found, the object reflected in the mirror plus the object on the paper forms the entire object again.

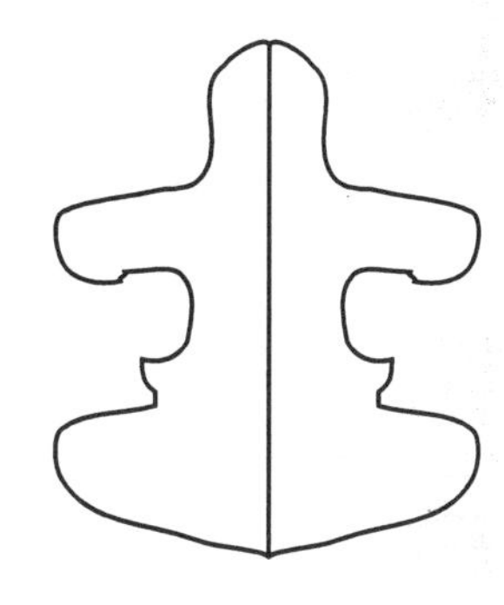

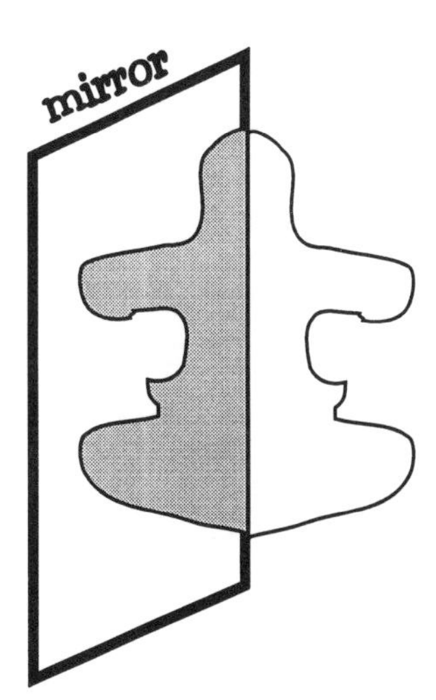

symmetry questions

? 1. Which of these objects have more than one line of symmetry?

? 2. Which have no lines of symmetry?

? 3. Whenever possible draw in the lines of symmetry.

Now let's go backwards. For each of these objects, suppose the dotted line is the line of symmetry, draw the other half of the object on the other side of the dotted line.

? 4. How many lines of symmetry can you find for these objects?

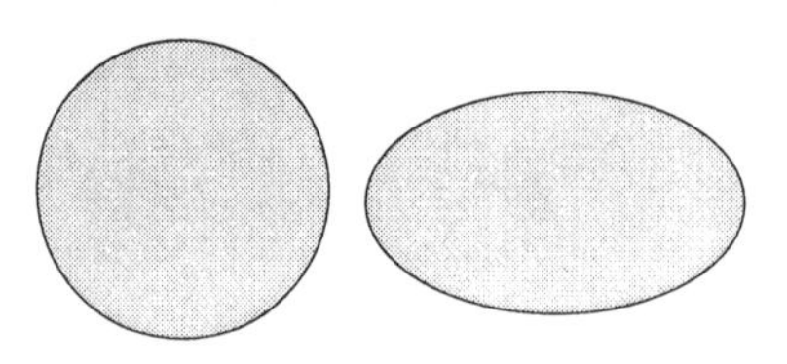

If an object has no end to the number of lines of symmetry through a certain point (in other words, it has infinitely many lines of symmetry that pass through a single point) then that object has *point symmetry.*

Place both your hands flat on the table in front of you.

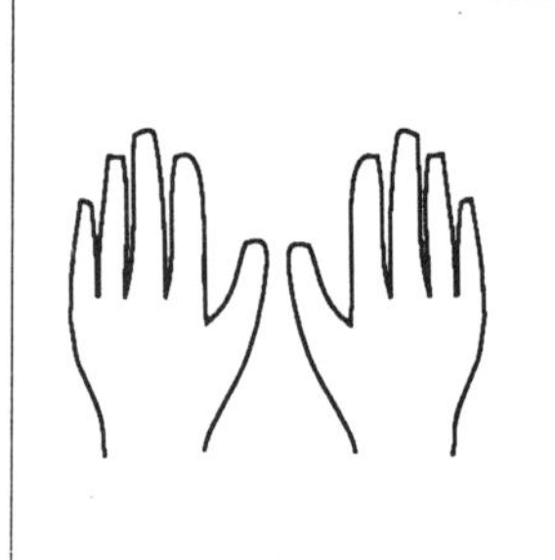

? 5. Do you see a line of symmetry here?

Now place both your palms together. Your two thumbs, two index fingers, two middle fingers, two ring fingers and two little fingers are together. A piece of paper placed between your two hands would act as a plane of symmetry for your hands.

? 6. Now here's the art problem – study the photograph of the sculpture done by the artist Auguste Rodin. Look at it very carefully. Something is very unusual about it. Can you figure out what it is?

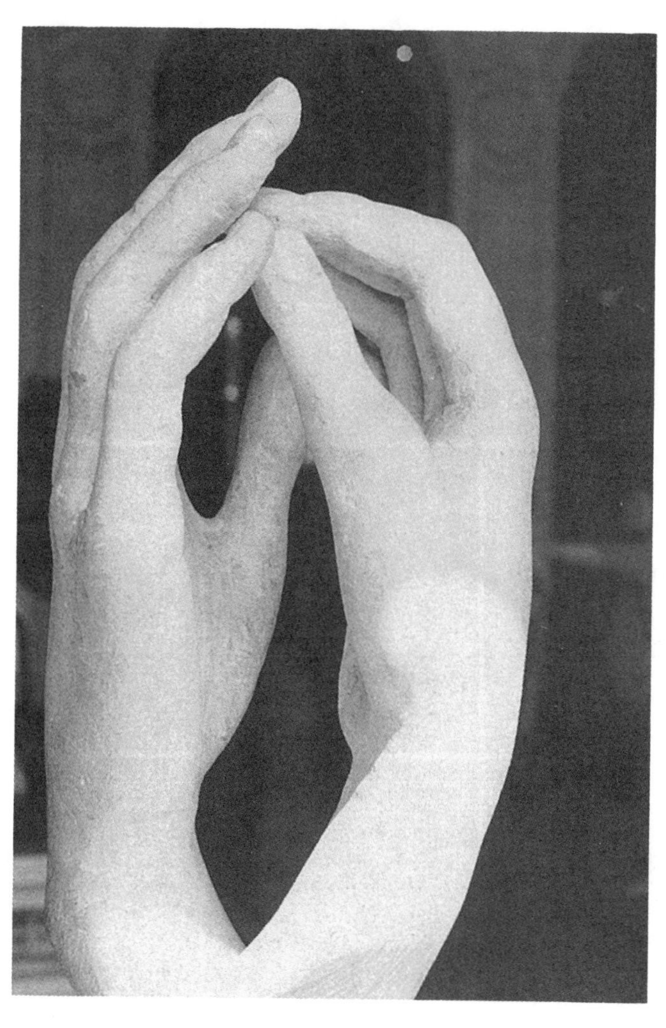

Answers to questions appear in the back of the book.

Factorials cut things down to size

It feels so good to be useful.

"It feels so good to be useful," ***period*** **said.** "In writing, I'm used to end a sentence, and in mathematics I am the decimal point. I separate the whole numbers from the decimal numbers. Yes, it feels *great*."

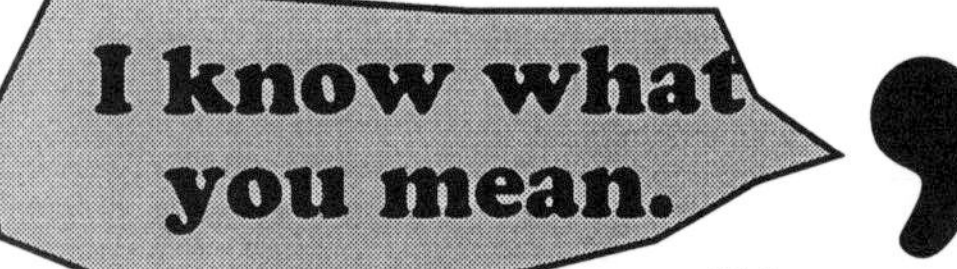

"I know what you mean," *comma* said. "I too am used both in writing and in mathematics."

"You both are so lucky. Most people know about you two, but not many people know about my mathematical use," *exclamation mark* said.

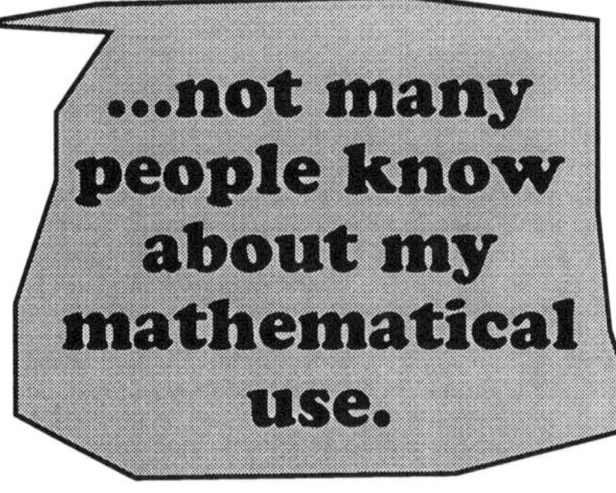

"Well, here is your chance, exclamation mark," said *period* and *comma.*

"What do you mean?" *exclamation* asked.

"You can explain to thousands of people in this book what else you do and why," the two replied.

"You are right!" *exclamation mark* said with a smile on its face. "Well, here *goes*."

"Mathematicians use many symbols to make what they write and do easier. For example, the symbol + means to add, and 5^3 means 5x5x5 or 125 – this is an example of three different ways to write the same quantity.

"Factorial is another symbol invented by mathematicians to make multiplication more compact. Notice the word *factor* in the word *factor*ial. Factorial uses the factors of a number. In mathematics, whenever you see **!** placed with a number like this, **9!** it means **9x8x7x6x5x4x3x2x1,** and it is no longer an exclamation mark. Answer the following questions and you will learn one way to explain how the factorial symbol came about."

• How many different ways can **three** different flags be arranged on a flag pole?

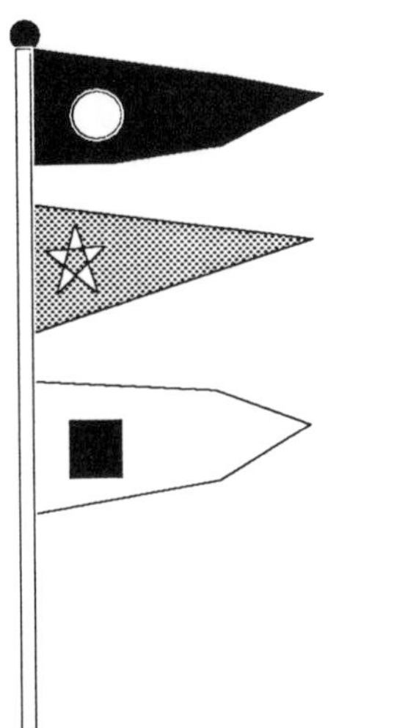

There are 6 ways.

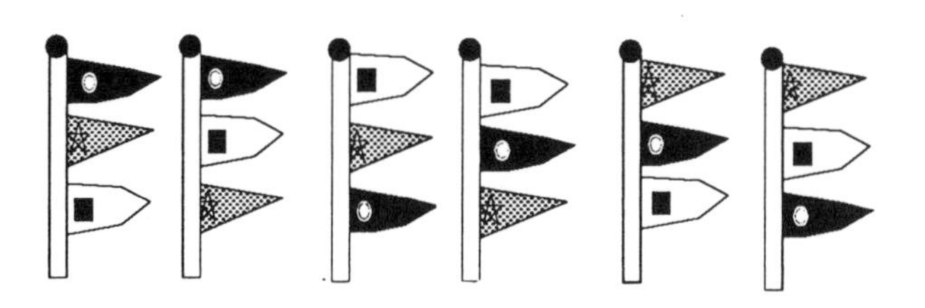

For the top position you have a choice of **three** possible flags. Once you selected one, that leaves a choice of **two** different flags for the middle position. And once you placed a flag here, that leaves only **one** choice for the bottom position. Now notice that **3 x 2 x 1 = 6.**

Now try the same thing with **four** different flags. We have a choice of **4** for the top, then **3** possibilities for the next, then **2** for the next and finally 1 for the bottom. **4 x3 x2 x1=24.** See if there are 24 different arrangements.

Now let's change the problem. Say **five** people rush to get in line for movie tickets.

• **How many different ways could they line up?**

Right! **5 x 4 x 3 x 2 x 1 = 120.** That is quite a lot.

If you had a hundred people, you'd get tired of writing out

100x 99x98x97x...

on and on until you get down to x 1. Well guess what? Mathematicians also got tired or writing these out. So what did they do? They invented a **new symbol** for shorthand. They used me **!.** So **5!** means **5x4x3x2x1.**"

factorial questions and experiements

The factorial symbol comes in handy with problems like these. It also is used in doing probability problems and other math problems.

1. Suppose I have 4 drawers. Each drawer contains 3 boxes. In each box there are 2 bags. Each bag contains 1 marble. How many marbles do I have? Can you write the answer as a factorial?

2. Some hand calculators have the factorial symbol on them. See if you can borrow one, and experiment with finding the factorial of a number using the calculator.

researching: Go to the library and check out *Anno's Mysterious Multiplying Jar* by Masaichiro and Mitsumasa Anno.

The rise & fall of the Roman numerals

So we're fat. We are still the best numbers around.

XV

The trumpets sounded.

Out marched the bulky and obese Roman Numerals. They had grown complacent, never seeking self improvement. They assumed their empire of Roman numerals would always be the numbers of first choice. They were convinced they would be victorious in any number encounter. Little did they know that a small rebellion was starting in a far off province to the East.

In the East new numerals had formed. These numerals referred to themselves as Hindu-Arabic. Their ten digits were

0123456789

They had a secret weapon which was far more powerful than all the forces of numbers of the Roman numerals. They had developed a new way of forming their numbers. A way so simple, yet very powerful, with which they were confident they would devastate the Roman numerals during a confrontation. You see, the Hindu-Arabic numerals had a special way of organizing their forces of numbers so they could outperform the Roman numerals in any arena—be it adding, subtracting, multiplying, dividing, etc. This new weapon was called place value.

The Roman numerals used the symbol **I** for *one*, **V** for *five*, **X** for *ten*, **L** for *fifty*, **C** for *one-hundred*, **D** for *five-hundred*, and **M** for *one-thousand* to write all their numbers. In other words, they repeatedly used these symbols to form larger and longer numbers by stringing them together. For example: *Two* was **II**, I+I. *Three* was **III**, I+I+I. *Four* was **IIII** (later **IV** which meant five minus one). *Thirty-eight* was **XXXVIII**, X+X+X+V+I+I+I. To write *eight hundred and forty three* got quite lengthy – **DCCCXXXXIII** which meant 500+100+100+100+10+10+10+10+1+1+1. But the Roman numerals didn't care if they were bulky and awkward. Can you imagine how difficult it must have been to add many columns of Roman numbers or multiply and divide with them? They did not have a place value. With the Hindu-Arabic numbers, it was the numerals location in the number that determined the size of the number. For example, **321** stood for *three-hundred and twenty-one,* 321. So adding meant just lining up the column of numbers, being careful to keep the ones, tens, hundreds, ... places all in line.

"The Roman numerals seemed to be stumbling over each other ... The Hindu-Arabic numerals had a secret weapon."

The merchants who went to far off lands to trade their goods and import new and exotic spices and fabrics, were the first to learn of these new numbers. Many of the foreign merchants they encountered would do their calculations very quickly and accurately using symbols and methods they had not seen before. The Arab merchant Esab Net became quite renowned among the traders for his skills in using these new numbers. He shared

0123456789

his knowledge freely with the Roman merchants, who were delighted to learn this new method of calculation.

* * *

The Roman Coliseum was buzzing with the news of the challengers to the official Roman number system.

The first test of power came when a Roman merchant presented the problem of adding LXII+CLVIII.

Esab Net did the problem quickly using the Hindu-Arabic numerals 63+158=221

The crowd was in an uproar. They couldn't believe their eyes. With each problem, the Roman numerals seemed to be stumbling over each other, solving the presented problem ever so slowly. The Hindu-Arabic numerals seemed to complete the problem almost instantaneously. And so the Roman numerals, with all their fanfare and bravura were overcome, and eventually the Hindu-Arabic numerals became the numbers of first choice.

* * *

The merchants gradually got hooked on using the Hindu-Arabic numerals. Their bookkeeping became so much easier. Now the revolution was in full force because this meant that merchants throughout Europe were slowly learning about and adopting this new system. They used it in their stores everyday with their customers. At first the customers did not want to learn new ways to compute, but it eventually caught on. Today the Hindu-Arabic numerals with their base ten place value system are what we use in our everyday arithmetic activities. Just think, if it were not for the evolution of these new numbers, you might be still adding and subtracting using Roman numerals.

The cycloid – an invisible curve of motion

There are many fascinating curves in mathematics. You probably know how to make a circle or

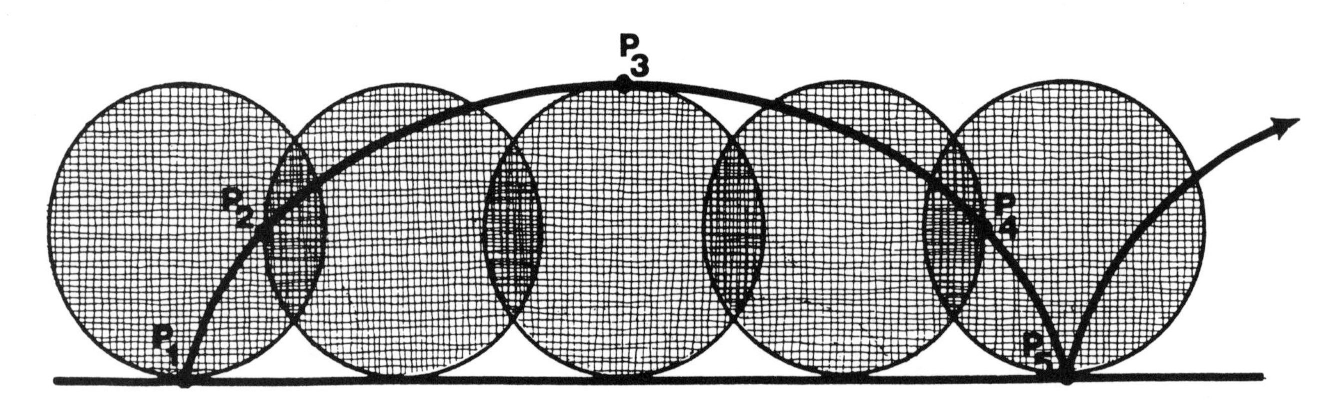

This figure shows the path a point travels on a circle which makes one complete revolution.

know what an ellipse looks like. But mathematics has many secrets. Sometimes you have to be a detective to discover some of its secrets. One such secret is the invisible curve of mathematics – the curve of motion. To find out what a cycloid looks like, we will need to do some mathematical sleuthing.

Go to your recycling bin and get a can. Attach a pencil inside the can by taping it down so the pencil's point extends beyond the can (see figure 1).

Now tape a sheet of paper on a wall next to the floor (see figure 2).

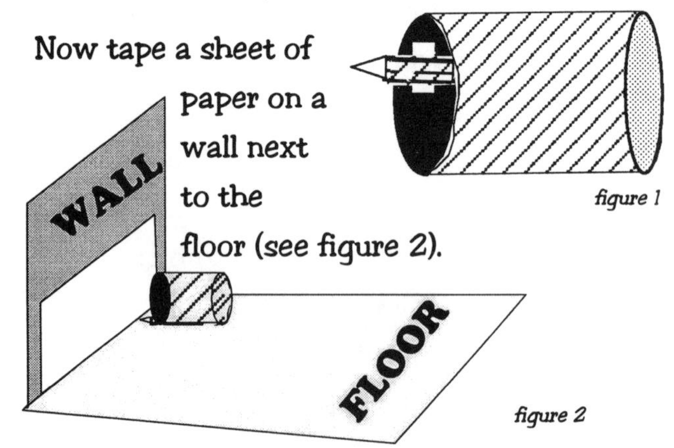

figure 1

figure 2

Place the can against the paper so the pencil point touches the paper. Starting with the pencil on the floor side, carefully roll the can against the wall so that the pencil traces a curve on the paper. Rotate the can completely three times. Did you get a curve that looks something like the cycloid?

This curve is the path formed by a single point (the pencil's point) as the rim of the can rotates. The curve made by this moving point is called a *cycloid.* The reason you normally can't see the cycloid is because *it is the path of a single moving point.* On the other hand, a circle is a curve made up of all possible points that are equally distant from its center. A circle is not the path of a moving point. Think of the air valve on the tire of a bicycle as a point. As you ride the bike, the valve is tracing out a cycloid's curve.

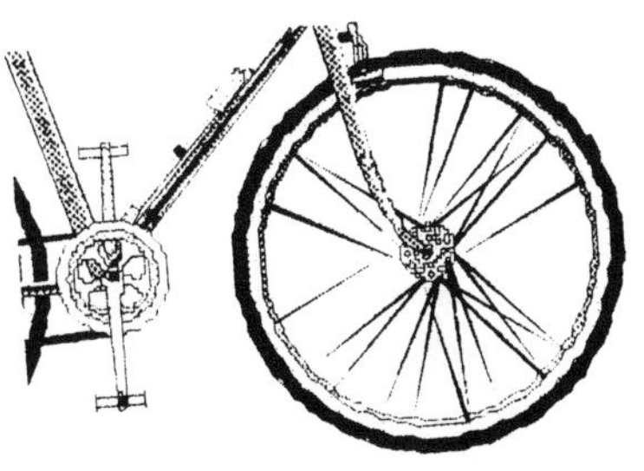

In the 17th century many interesting discoveries were made about the cycloid. You can discover one of its properties by taking measurements off the path of the cycloid you just made.

Experiment

Get a long piece of string and a ruler. Using the string, measure the arc of your cycloid by carefully draping the string along the cycloid's arc. Now, using the ruler, measure the length of this string. This will be the length of your cycloid's arc. Now measure the diameter of the point's rotating circle by measuring the diameter of the can's rim from pencil point to pencil point. *How many times longer is the cycloid's arc than the circle's diameter?*

Right! Four times longer. A cycloid's arc will be four times the size of its rotating circle.

Cycloids appear in many places you may not expect. Cycloid curves appear in ocean waves, because these waves involve circling particles. Even a train's wheel makes a special type of cycloid shape, as in this diagram.

You can be sure, wherever you see a rotating circle, an invisible cycloid curve is being formed.

When the operators came into town

Thousands of years ago numbers had it so easy.

There was a time when people used only counting numbers. For example, ||||| could be used to show five fish had been caught. Counting numbers were the only ones who inhabited Numberville at this time.

the operators

+ − × ÷ √

"I was sent here to operate on you."

+

The problems that required numbers were just counting problems. So the counting numbers had a very easy and lazy life. In fact, those less than twenty were the most overworked because people did not have a whole lot of things to count or keep track of.

"This is the life," 1 said as it lay sun bathing near the pool. "I like being a number. The work is rewarding– helping people keep track of their things." "And there is plenty of leisure time," 4 added, while sipping on a cool beverage. Other numbers were playing volleyball. Some were called on to work for a few moments, but soon returned to their idyllic life in Numberville.

"I don't know what you're talking about."

1

As time passed, people made more and more demands on the counting numbers, but things were still very mellow.

Then one day some strangers came into town. They were things that no number had ever seen before. The one that looked like a big cross was the first to speak. "I am looking for the number 1," the + asked the number 7.

"Well, you should find 1 on the patio at this time of day," 7 replied, a bit hesitantly.

The + went to the patio , and immediately spotted 1. "I have been looking for you, 1," + said. "I have work for you."

"Who are you, and what work could you have for me?" 1 asked cockily.

"I am an operator. I've been sent here to operate on you," the + replied emphatically.

"I don't know what you're talking about," 1 said. "I don't know you."

The + became angry and grabbed 1 and the 5 that was nearby and joined them, 1+ 5. The + didn't stop with this. The + **operated** with 1 and every number it found at the patio, so there were 1+ 8 and 1+ 4 and 1+ 12 and 1+ 9.

"Stop," shouted 1. "I am getting tired."

"We are just getting started," the + replied. "Wait until you meet the other operators."

"Other operators?" 1 repeated with a fearful note in its voice.

"Sure, there are lots of us. There are - and **x** and ÷ and squaring and square rooting and lots more are on the way."

"I don't like being tied up with all these other numbers. It is not relaxing. Please release me," 1 pleaded.

"No way. Our job is to perform operations. And once we have made an expression, there is no way we'll let you out."

The operators held the numbers of Numberville captive. They grabbed numbers right and left and operated on them. 3 **x** 2 and 12÷4 and 23+ 4 and 7- 5 and 1 **x** 1 and 1÷1 and 9 **x** 2 and on and on. There was no rest for the numbers. Apparently people had discovered how to operate with numbers. They did not just use them for easy counting, but instead for more complicated arithmetic. It seemed that the days of leisure in Numberville had come to an end.

Fortunately, the number 2 had not spent all its days lounging around the pool. It had been studying the discoveries that people had been making, and had known that the operators were in the offing. 2 believed it knew a way to free the numbers from their operators, whenever they wanted to be released. 2 decided to try it out on the 2×3 operation the **x**

+ “Stop this.”

had done to 2 and 3. 2 had read that every operation has an inverse operation, one that does the opposite operation. 2 had learned that the inverse operation for **x** was ÷. So 2 tried it. It took $(2 \times 3) \div 3$ and sure enough the result was 2. It tried it with $12 \div 4$ using **x**. $(12 \div 4) \times 4$ gives 12. Thrilled by what it had found, 2 dashed

over to 1 and explained its discovery.

1 was overwhelmed with joy. "I can release myself from 5 just like this– $(1 + 5) - 5$ gives 1. Hoorah!"

"Stop this, the **+** shouted. You are not supposed to know about our inverses. You can't undo our work."

"We can if we want. You are no longer in control. We know your secret. You must have our cooperation, if you want to operate." 1 replied.

"What good are operators without numbers?" **+** asked. "We need you." Now it was the **+** who was pleading.

"My life was too boring before you arrived," 2 spoke up. "I did not like relaxing most of the day. And I am sure the other counting numbers will admit this, if they think about it realistically. Can't we make all our lives more interesting by working together?" 2 asked.

" 2 is right! " the counting numbers shouted together.

"Well then, let's CO-OPERATE together," the operators joyfully replied.

Figurative numbers & ants

Today, Suzy just did not feel like doing homework. Her mind was on other things. She wanted to ride her new bike and show it off to her friends. As she sat daydreaming she saw a stream of black dots marching along the left edge of her desk. "Oh no! ants," she thought, but she did not feel like doing anything about them. Instead, she just stared at the moving trail. To her amazement the ants seemed to be crawling in formation. It was as if the top of her desk had become a football field and the ants were moving like a marching band at half time. She was fascinated by the formations they were making.

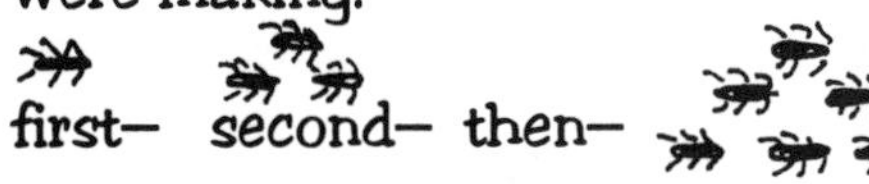

She had never seen ants do this before. Yet, on the other hand, she had never watched ants so closely or for so long.

All of a sudden the head ant shouted to her, "Well, can you guess?"

"Guess what?" She asked, a bit startled to be talking to an ant.

"What we are showing you," replied the ant.

"Showing me?" she questioned, with a confused tone to her voice. "You're JUST making interesting figures," she said.

"More than that," the ant replied. "We are from the mathematical corps of marching ants. We're demonstrating our newest routine using figurative numbers. See if you can determine what they are and why they are called figurative numbers."

"Figurative numbers," she thought, "I just read about them in my math assignment." As Suzy watched the ants and their formations, she realized what these numbers were all about. She quickly jotted down what she had discovered on her homework paper. "You ants can practice as long as you want on my desk," she said happily, as she left to dash off on her new bike.

?? questions ??

How did Suzy answer the following homework questions? *HINT: Look at the diagram showing the ant formations on page 27.*

? 1) What is the value of each of the triangular numbers?

Without drawing the 8th triangular diagram, how many dots will be in the pattern?

What is the 8th triangular number?

? 2) 1, 4, 9, 16, 25, ... these are square numbers. What diagrams do they make?

Is the number 49 a square number?

Which square number is this?

Now draw the square pattern for the 8th square number.

Answers are given in the solutions sections at the back of the book.

Can you find the first eight odd numbers – 1, 3 ,5, 7, 9, 11, 13 and 15 –in this square?

Also notice that the sum of the first eight odd numbers is 64: 1+3+5+7+9+11+13+15=64

Does this idea work for all square numbers?

? **3.** Here are some of the names for some figurative numbers–

triangular

square

pentagonal

hexgonal

Can you draw the diagrams for the first four pentagonal numbers?

See the end of the book for answers to these questions.

Pythagorean triplets

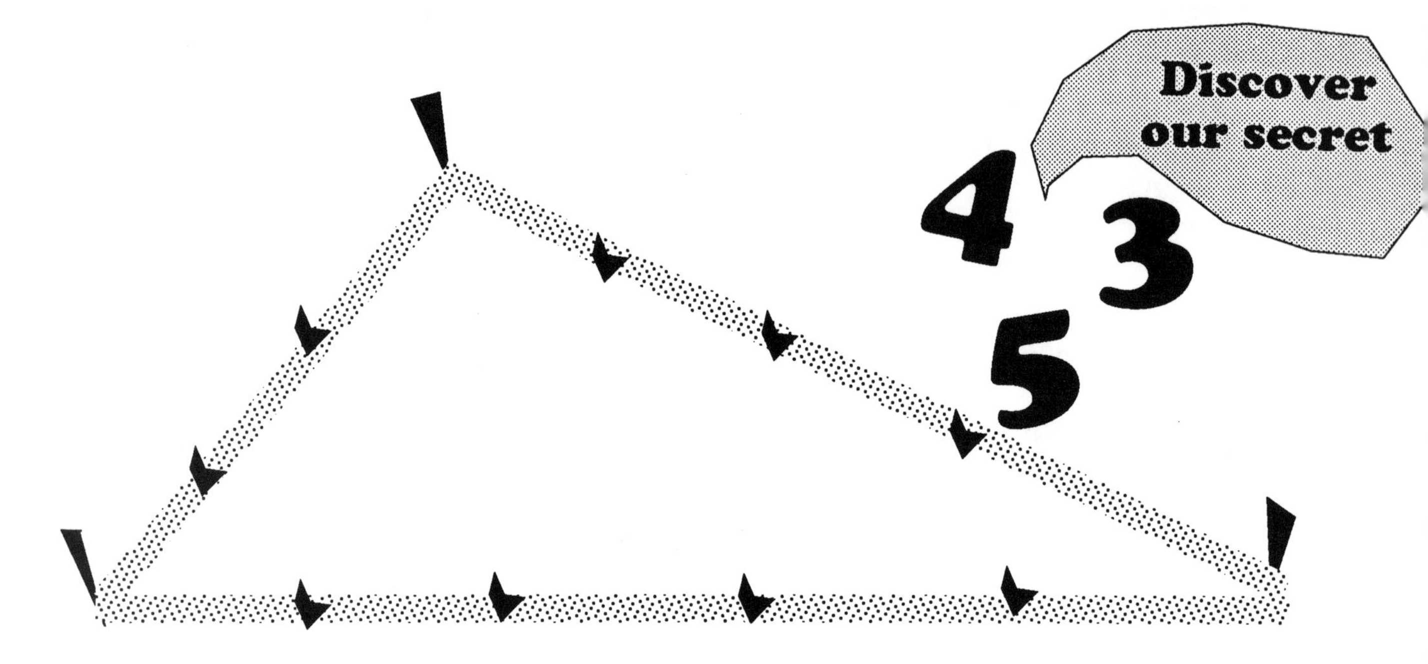

Lately, 3, 4 and 5 always seemed to be hanging around together.

And it wasn't just because they were three consecutive whole numbers. They were conspiring, and no other numbers knew what they were up to ... until they called on 12 and 13.

* * *

1 decided to find out what was up. Number 1 invited 3, 4, 5, 12 and 13 to have tea with it. As they were enjoying their tea, 1 began "I've asked you over because lately you have been acting rather elitist ."

"That is because we are on a secret and very exciting mission of discovery. You will learn about it soon enough," 5 responded snobbishly.

"I already know your secret," 1 said, with a smile on its face.

"How can you know? Who let out the secret?" 3 asked, looking at the other numbers.

* * *

To discover their secret, get a ruler and draw a triangle with sides 3", 4" and 5". Do the same with 5, 12, and 13.

Do you notice a special characteristic that both triangles have?

"For thousands of years, I have known that {3, 4, 5,} and {5, 12, 13} are numbers which form right triangles – triangles with a 90° angle. You probably do not know that every triplet of numbers that makes such a triangle is called a *Pythagorean triplet.*" 1 said.

"We are?" 5 asked with a startled look on its face.

"How many triplets do you think there are?" 1 asked.

" Just us, {3,4,5} and {5,12,13}," 12 replied.

"No, no. no!" 1 said. "I'm the oldest of numbers and probably the wisest. So let me tell you that there are *infinitely many* Pythagorean triplets." A startled look came across the faces of 3, 4, 5, 12 and 13.

"We thought we were members of a small elite group. We had no idea so many special triplets existed," 13 said.

"Why should you be disappointed or secretive? I could have told you that Pythagorean triplets have been around for centuries. The Babylonians and the Egyptians used them before the Greeks," 1 explained.

"But why would someone want to make right triangles?" 5 asked.

"There are not many houses or buildings that don't have right angles. Look at the corners of this book. 90°, correct?" 1 said. "Take any rectangle or square. They each have four right angles. Fold one along its diagonal. What do you get?" 1 continued. "You get right triangles. They are all around you– tables, chairs, fences, football fields, basketball courts, desks. Pythagorean triplets were the only way the ancient Egyptians had to make right angles for the various buildings and structures they constructed," 1 explained. "They first made the right triangle by using ropes the lengths of a Pythagorean triplet."

"You should still be proud you are

Pythagorean triplets, even though there are countless others. You all belong to a very special group," 1 said encouragingly.

"How do we find some of the other Pythagorean triplets?" 4 asked. "That's easy," 1 said. If you double yourselves – 3, 4, 5 – you get 6, 8, 10. It's is also a Pythagorean triplet. In fact, any multiple of 3, 4, 5 is a Pythagorean triplet. The same works for any multiple of 5, 12, 13," 1 explained. "Try it!"

making a discovery

1) Which triangles below look like right triangles?

1 2 3 4 5

2) For each of these Pythagorean triplets – {3, 4, 5} or {5, 12, 13} – give three other triplets which you get by multiplying the numbers of each triplet by:

3 ________ ________

8 ________ ________

10 ________ ________

3) By what would you multiply the Pythagorean triplet {3, 4, 5} to get the Pythagorean triplet {3/5, 4/5, 1}.

researching:

• Over the centuries mathematicians have devised formulas for finding other Pythagorean triplets, you might want to look up some of these in the library or perhaps on the internet.

Check your answers with those at the back of the book.

Pythagorean triplet experiments

90°

1) Using a ruler make a right triangle by extending the sides around the 90° angle shown in the following way:

- *Extend one side to 1 3/4 inches and the other to 6 inches.*
- *Connect the ends of these two sides to complete the right triangle.*

How long is the 3rd side? ______

Take the numbers of this triplet and multiply them each by 4. What numbers did you get?

______ ______ ______

This is also a Pythagoren triplet. These number will also form a right triangle. Try it.

2. Using your ruler make a triangle with sides of length 3", 3" 4". Is this a right triangle.

Answers are given at the back of the book.

Cutting up mathematics

Get a pair of scissors, a pencil, and some scrap paper. We are going to cut up mathematical objects.

On a piece of scrap paper make a copy of this shape.

Using scissors, can you make one straight cut and get two pieces that form into a triangle?

How about copying this triangle and cutting it into two pieces that fit into a rectangle?

Now copy this triangle. With one straight cut get two pieces that form a square.

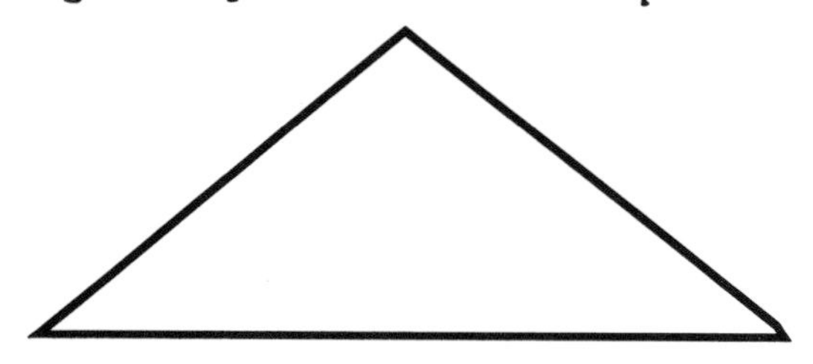

How about cutting this curving object with one straight cut, and end up with two pieces that fit to form a square?

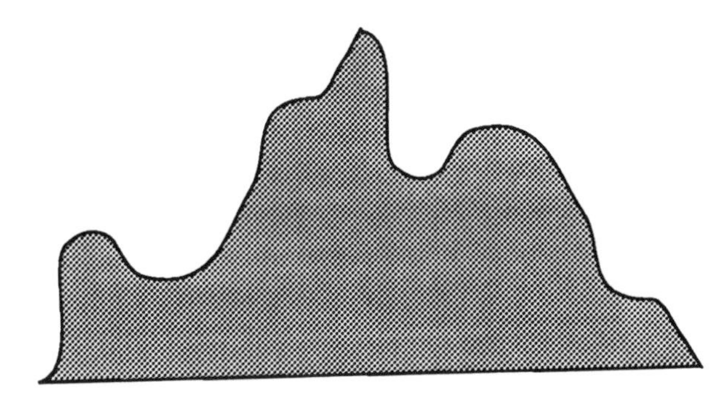

These are called dissection problems. You can make up your own. In fact, the famous mathematician David Hilbert proved that any polygon can be transformed into another polygon of the same area by cutting it up into a finite number of pieces.

One of the most famous dissection puzzles was created by the renowned puzzlist Henry Ernest Dudeney. He took a triangle and cut it into four pieces so that the four pieces formed a square. Here is how he cut up the triangle.

Can you figure out how to put them together to get this square?

Once you have done this, there is an explanation at the back of the book that shows how to attach the four pieces of the triangle with a string so that the triangle and the square can be changed easily into one another. The string holds the pieces together in the right order so that one form rolls into the other.

Solutions are given at the end of the book.

a challenge:

Make up some of your own dissection problems.

Paradoxes tease the mind

In a particular village in the Alps, the barber shaves all those in the village who do not shave themselves. Who shaves the barber?

The Island of Baras was situated smack in the center of all the shipping routes. It had become a wild place, a haven for pirates. Thieves, pirates and murderers did whatever they pleased, and Baras was now a lawless mess. That is until Ricco became governor. Ricco himself had once been a pirate. He knew what needed to be done to put things in order. During his campaign he promised that if he were elected things would change. Once again Baras would become a quiet sleepy island where people could leave their doors open and not fear robbers or other intruders. Needless to say, with such promises, Ricco was elected. Ricco enacted many strict laws. For example, anyone caught stealing would have to make good to the islander fourfold. Anyone coming to Baras had to state why they were coming. If they told the truth, they could go about the island freely, so long as they did not bother anyone. If they lied, they would be hanged. These were the laws Ricco had established, and everyone followed. Things were going smoothly, and peace was restored to the island.

One day an unusual traveler arrived. He had a large beard, big brown eyes, and wore a strange looking pointed hat. "Why have you come to Baras?" Ricco asked him. The traveler paused for a moment and replied, "I am here to be hanged." What could Ricco do with such a reply? *If the man was telling the truth he must go free, but if he went free he was not telling the truth because he said he was there to be hanged. If he was hanged, he was telling the truth, and must go free.* WHAT A DILEMMA!

Such a situation is called a ***paradox.*** It is something with a contradiction in it. A paradox seems to make sense yet at the same time does not. Paradoxes come in all forms—stories such as this one, diagrams, sentence, mathematical forms. Here are some of Penrose's favorite paradoxes, can you find the contradiction in each?

Do not read this sentence.

See solutions section for additional explanations of some of the paradoxes presented.

The infinite sets paradox

The counting numbers are the numbers 1, 2, 3, 4, 5, 6, 7,... They never stop. They are an infinite set. The even counting numbers are 2, 4, 6, 8, ... They also are an infinite set. Each counting number can be matched with a even counting number in the following way.

{1, 2, 3, 4, 5, 6, 7,...}
{2, 4, 6, 8, 10, 12, 14...}

So each counting number is matched with an even number twice its size. Since both sets are infinite, the matching will never stop. So there must be the same number of counting numbers as even numbers. But how can this be, since the counting numbers are composed both of the infinite set of even numbers and the infinite set of odd numbers?

The Impossible figure paradox

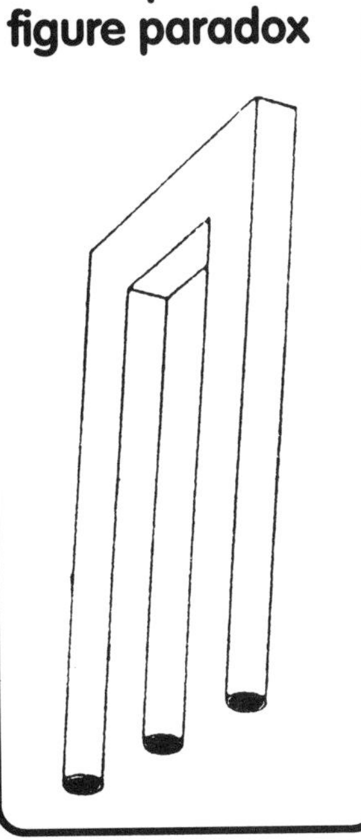

The missing bar paradox

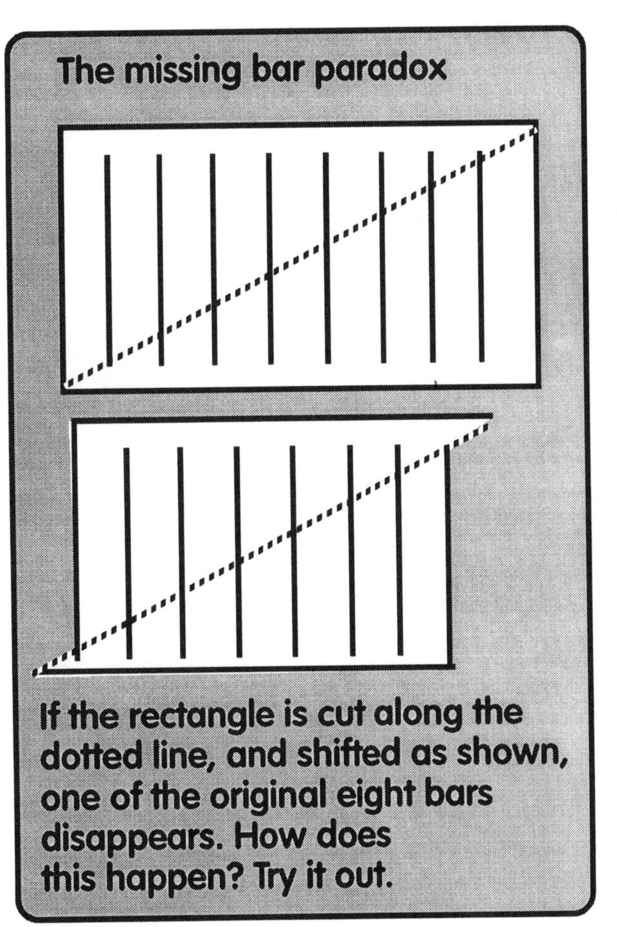

If the rectangle is cut along the dotted line, and shifted as shown, one of the original eight bars disappears. How does this happen? Try it out.

The day the counting number split up

"Who ever counts by odd numbers?" sneered 2. "You never hear of it," he continued to tease the odd numbers. "The rhyme is all off. Counting by 2s is best– 2, 4, 6, 8, … ."

"Wait a minute," shouted 5. "Counting by fives is just as good if not better—5, 10,

15, 20 .... It has a nice rhythm to it. And besides, there are five sticks of gum in a pack."

"You silly numbers," yelled 10. "Ten is the superior number to count by—10, 20, 30, 40, ... —tens are the easiest to work with."

"No," contested 3. "By 3s— it's more intriguing. I'm the mystical number. Everyone knows things always happen in 3s."

Each number protested and thought itself better than the other. Back and forth they continued. Finally, 2 said—"I'm taking myself and all my multiples with me and leaving the counting numbers."

"Wait just one minute," shouted 3. "You can't do that."

"Yes I can," 2 said. "Why not?"

"Some of your multiples are my multiples. Have you forgotten them? You can't take 6, 12, 18, and on and on, " 3 said.

Then 10 jumped back into the discussion. "All my multiples are your multiples. They are all even—10, 20, 30,... You cannot drag me and all my multiples off because you dec—"

Suddenly a deep loud voice they all recognized interrupted. "Stop this bickering," said the SET OF COUNTING NUMBERS. "None of you can leave. As a unit we are a set, a famous set—the counting numbers. We were the first numbers people used for counting. None of you can be taken away and expect the counting numbers to do their job—to count. Sure, each of you forms subsets in your own right, but you are interrelated. 2 and its multiples cannot walk away if the multiples of 10 and the even multiples of 3 don't want to."

Realizing the numbers understood its point, the SET OF COUNTING NUMBERS continued in a gentler tone, "Let's get back in order. I am taking roll call. We have to be ready to be used at a moment's notice. {1, 2, 3, 4, 5, ... }."

questions on multiples

? 1. Which numbers below do not belong to the set of multiples of 3?

23 24 15
31 30 6

? 2. Which numbers below belong to the set of multiples of 10?

30 15 50 25 100

? 3. List the multiples of 4 which are less than 32.

? 4. List the multiples of 5 which are between 15 and 35.

? 5. What number is the smallest multiple of 12?

? 6. Which number is the largest multiple of 5 less than 200.

The day the solids lost their shapes

The solids were beginning to be such a pain.

They were always going around bragging about their dimensions. But today would change all that.

* * *

"Doesn't it feel great to be above things like points, squares and triangles?" the cube asked the square-based pyramid.

"It sure does. We are so superior to any triangle or square," the pyramid added.

"Wait just one minute," shouted a square. "What do you think you're made up of but six squares."

"Yes, but I have more dimensions to me," the cube replied.

"That is true, but I can easily bring you down to my level." With that the square grabbed two of the cube's vertices.

"Wait a minute. I'm slipping," screamed the cube as it slid flat down next to the square, and ended up looking like this.

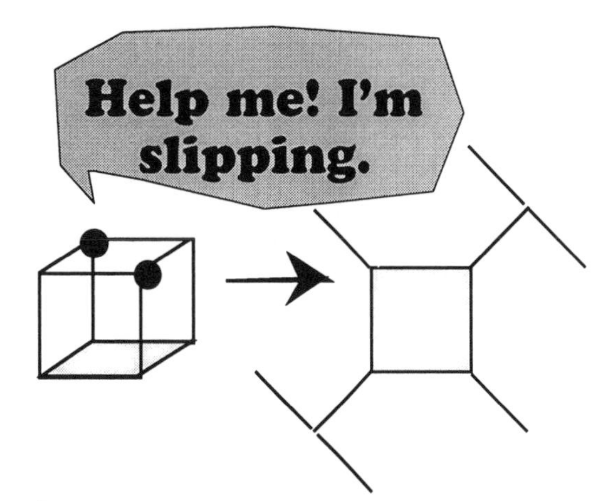

"See, you're no longer 3-dimensional," the square laughed. "I must say you make an interesting pattern. Let's see what flat patterns the rest of you make?"

* * *

Help the square and the triangle discover the flat forms of the solids shown. You'll need to visualize how the solids will look once they are collapsed. The round black vertices are the ones that are to be removed. Then let the solid fall flatly apart. Notice the same solid collapses in different patterns depending on which vertices are removed.

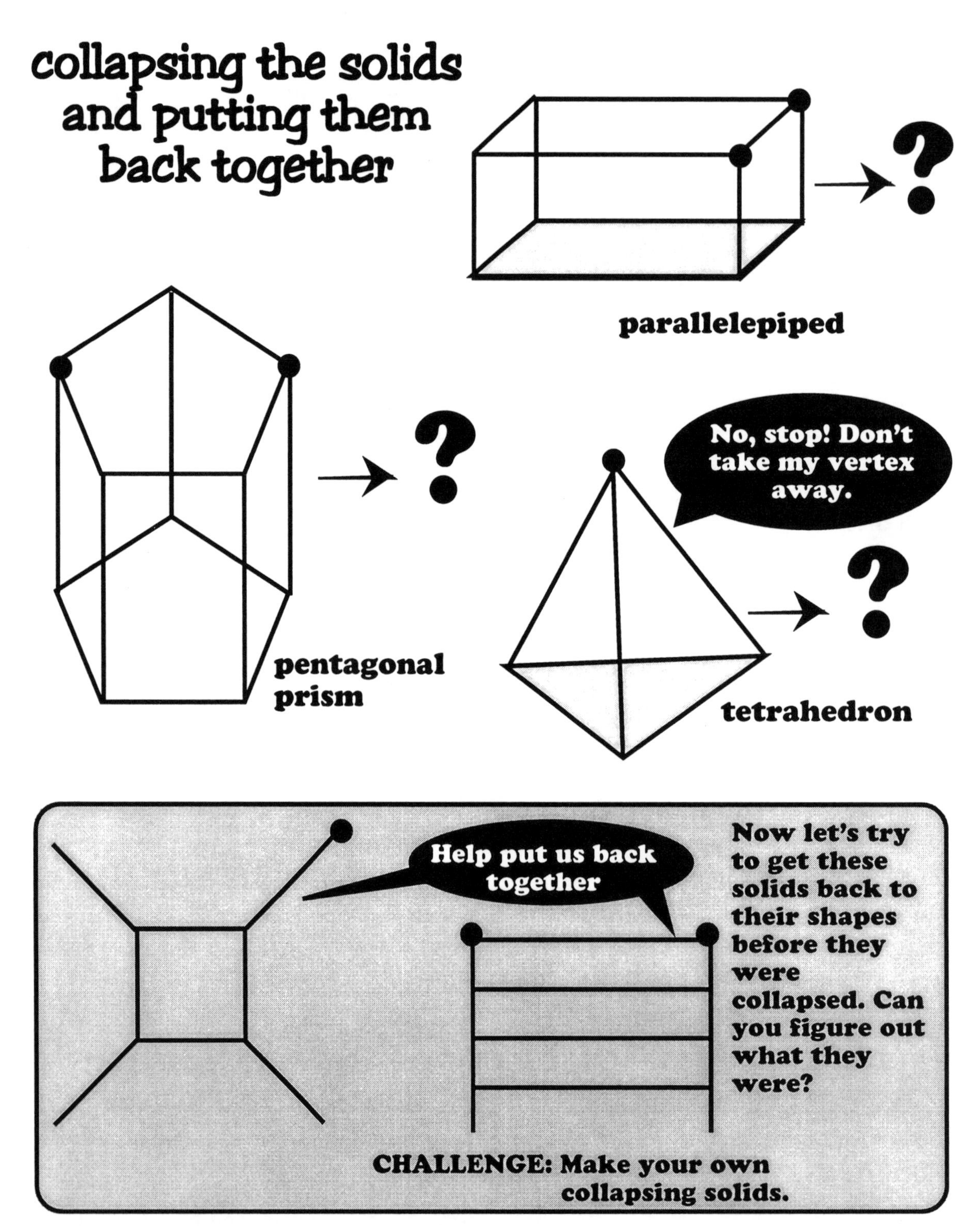

Solutions are at the end of the book.

The masquerade party

Fractions are a wild set of numbers.

There are proper fractions, mixed fractions, improper fractions, reduced fractions. There can be infinitely many fraction names for the same number. For example, 2/3 can be written as 4/6 or

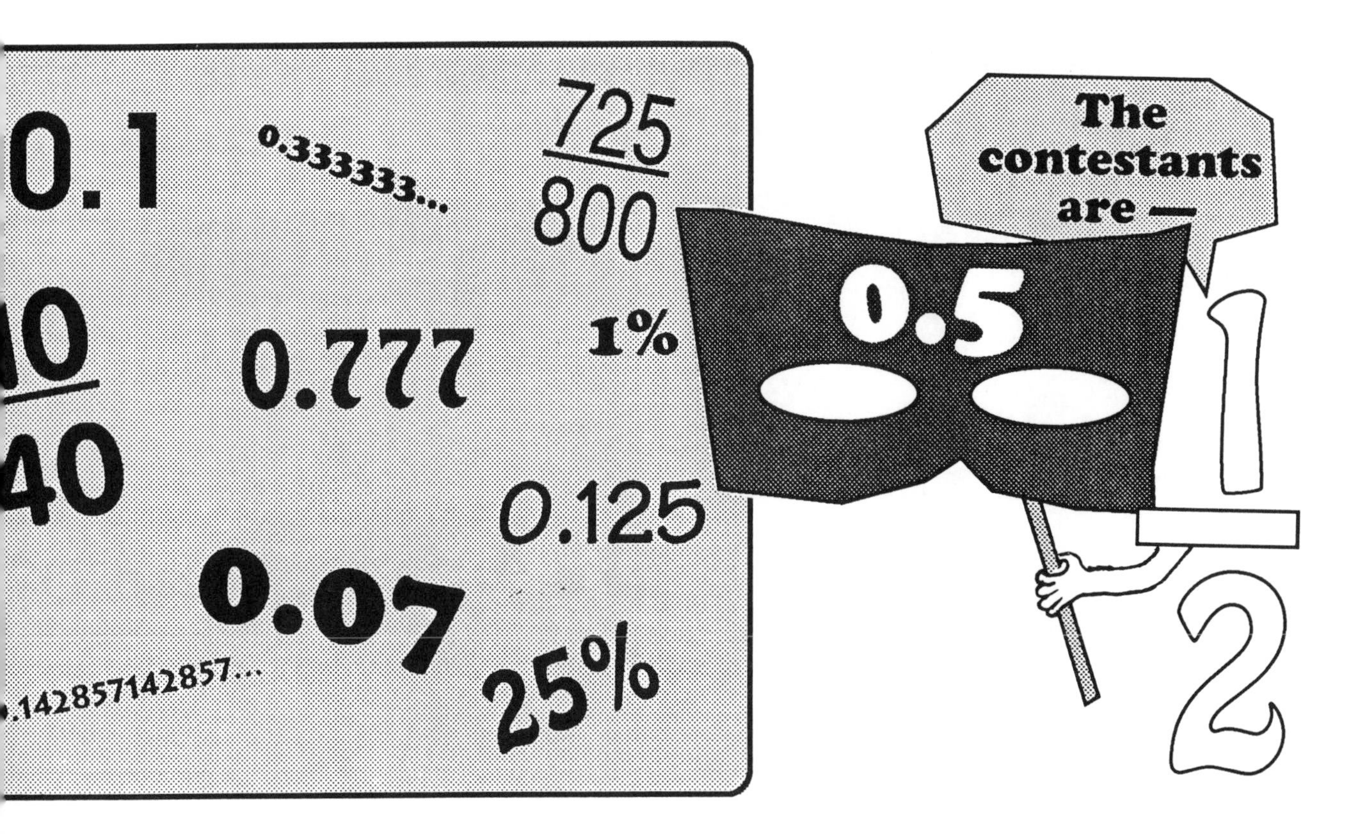

6/9 or 8/12 or 10/15, and on and on because when these fractions are reduced to lowest terms the result is 2/3. Since there are so many ways to express fractions and numbers, the fractions thought it would be fun to have a masquerade party. The party was open only to fractions, but each had to wear a costume to disguise its reduced fraction form.

Fractions arrived from all over in outfits you wouldn't believe. 0.3333333333333333... walked in carrying an infinite tail of threes, which it sashayed through the dance hall. 10/40 walked over to 0.3333333333333333... and said, "You're awfully cute, but I don't recognize you."

You're awfully cute, but I don't recognize you.

10/40

"That's the whole idea, " 0.3333333333333333... replied.

"Would you like to dance," 10/40 asked, and off they went.

At a table near the orchestra there were fractions with various masks. 25% sat next to 0.125 and 0.1 who were talking with 25/50. Underneath each of their masks was a fraction in reduced form. The ballroom was full of these crazy costumed fractions.

Suddenly the orchestra played a fanfare, and all the partygoers turned toward the stage. "It's so good to see so many of you having such a great time, " 0.5 said. " It is getting close to the time to judge the best costume,"

That's the whole idea.

0.3333...

0.5 continued. "I would now like to ask all of you fractions entering the competition to come up on stage."

"Here they are," 0.5 declared as it presented them.

0.1 25% 0.142857142857...
0.3333333333333333... 0.125
10/40 725/800 1% 0.777
0.07

"As agreed the winner will be the reduced fraction whose costume is so good that its reduced form cannot be identified. Remember it is not just finding a fraction, it must be in lowest reduced terms, where its numerator and its denominator cannot be reduced further. May the best masquerade win! "

* * *

You are the judge. The winner will be the fraction you cannot identify. If you identify them all, you are the winner. The fraction forms are listed in the solution section at the end of the book.

Check the solutions section at the back of the book for the fractions represented in the contest.

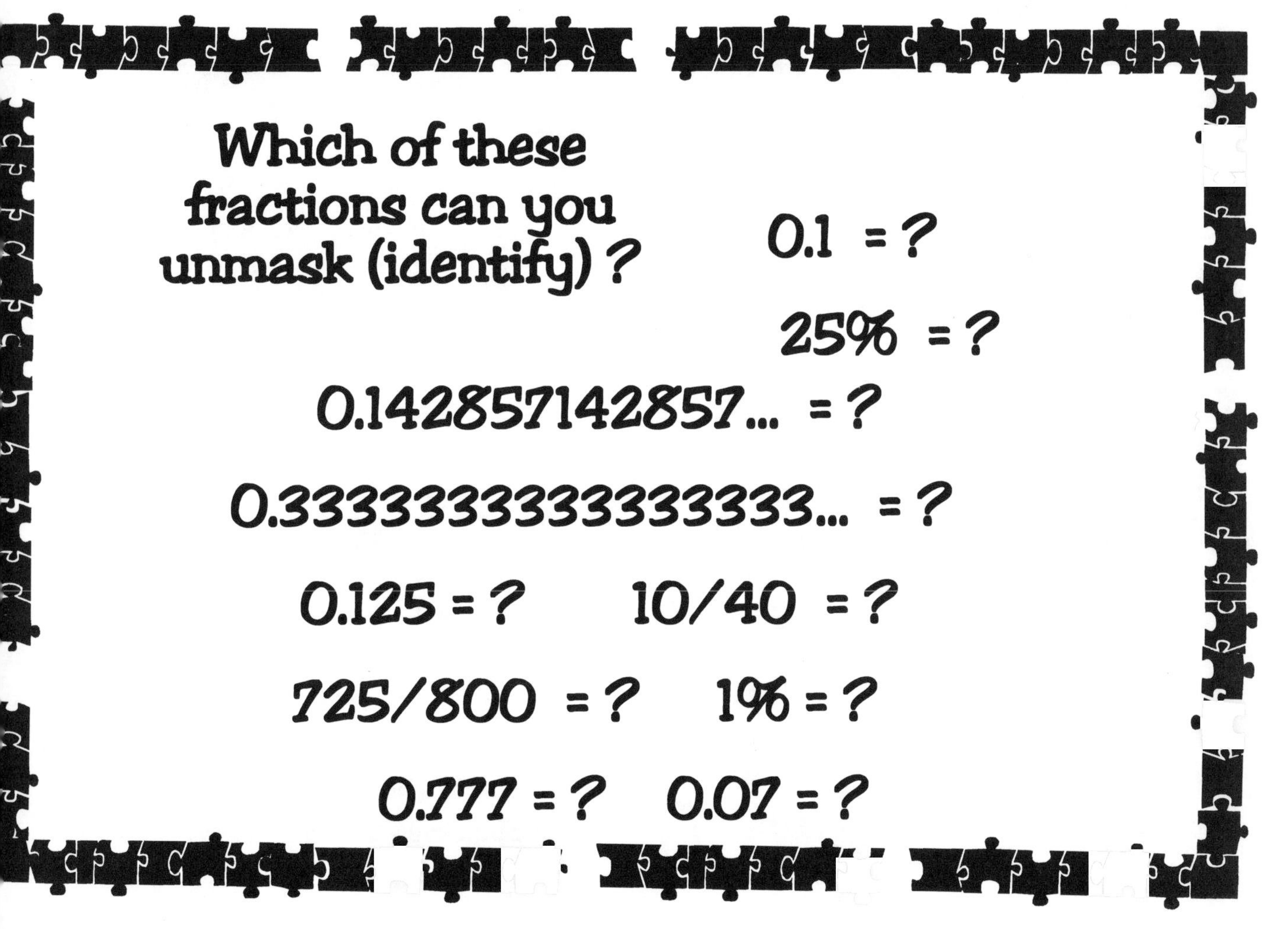

The Persian horses

The story of the painting of the Persian horses dates back many, many centuries.

Optia was known in the ancient world as an artistic magician. Her magic was very different from that of other magicians. She painted magical pictures. Pictures which would change as you looked at them. Pictures that seemed to hide things which would appear as you stared at them. Pictures that danced before your eyes. Pictures that could easily deceive you.

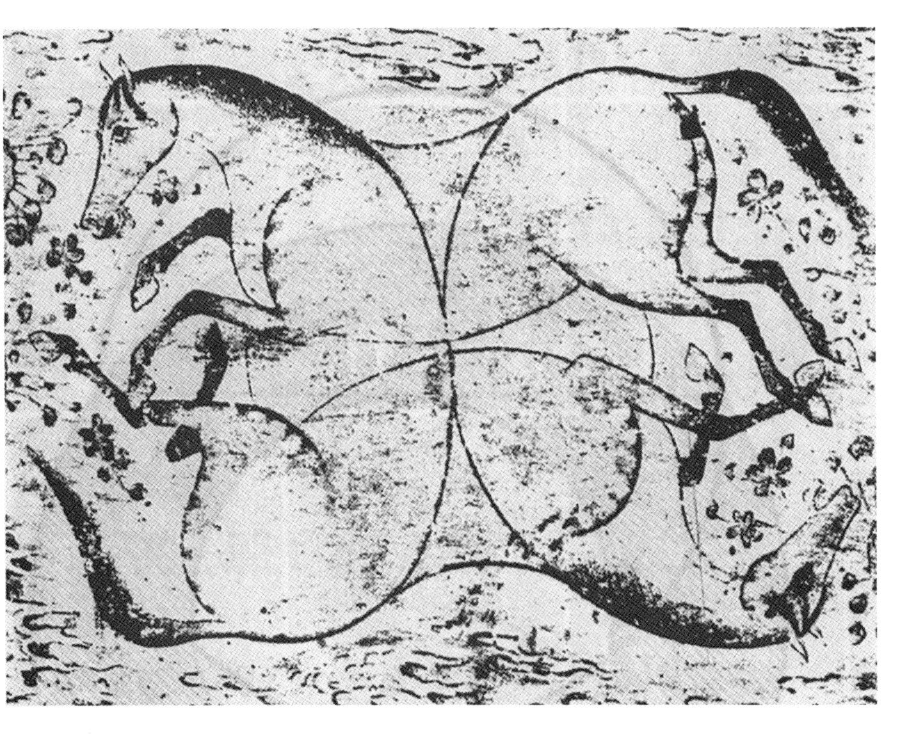

How had Optia learned to paint such treasures? Her teacher had been an ancient man who had learned the secrets of numbers and figures. He had discovered how to use shapes to create illusions. The old man had recognized her artistic talent, and wanted to pass his knowledge of how to create optical illusions on to Optia. Optia was an exceptional student. She learned quickly and appreciated what the old master taught her. In graditude, she watched over him as long as he lived.

Using what she had learned, Optia created wonderful and unusual works of art. Eventually her work was in great demand and even the King of Persia asked her to do a painting of his horses.

Optia knew the king's kingdom had once been divided into two regions. Later, each of these regions had split into two parts. Instead of painting just the king's horses, she painted the Persian horses so they symbolized his kingdom. The king waited anxiously for the painting to be completed. When she had completed it, she brought it to the king. When she unveiled her work, the king complimented her on her fine artwork. Suddenly the king's eyes opened very wide. It was as though something new had appeared before him. And sure enough, something new had appeared. The two horses now appeared to be four. He was amazed! One moment he would see two horses, and the next there were four—symbolizing the parts of the kingdom. The king was thrilled with his painting, and he gave it a prominent place in his palace. And Optia's fame increased tenfold.

Here are some special optical illusions techniques Optia learned from her teacher. Perhaps you can use some of these techniques and figures to create your own drawings.

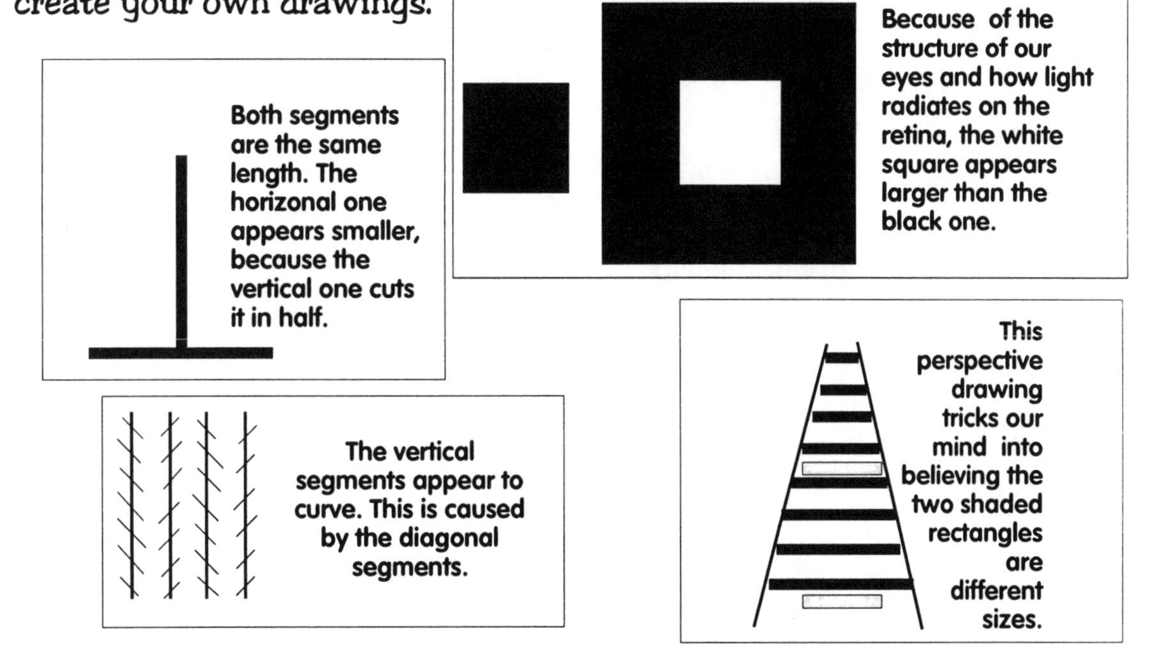

Welcome to MEET THE FAMOUS OBJECTS show

Welcome to the MEET THE FAMOUS OBJECTS show. Today we are interviewing that famous multi-personalities figure—the triangle.

Host: In the past we have interviewed squares and rectangles, but we have never had you on our show before. What special features makes you such a prominent guest?

Well, as you know a square is a square—four equal sides and four right angles. It really does not change its format. The triangle has the minimum numbers of sides for a two dimensional object. All other two dimensional objects have more than three sides. But when you say a triangle, you are also talking about a host of different objects.

Host: Like what?

Like isosceles triangles, equilateral triangles, equiangular triangles, scalene triangles, right triangles,.........

Host: Wait just a minute. You're throwing too many new words at the viewers. We will need you to give explanations of these figures. Better yet, how about giving us illustrations of these objects?

No problem.

An isosceles triangle has two sides the same length. For example—

An equilateral triangle has all its sides the same size, and happens to also be equiangular, meaning all three angles are the same size, 60° each.

Scalene triangles have no parts the same size, so their sides and angles are all different.

A right triangle is one that has a right angle, namely a 90° angle. Like this—

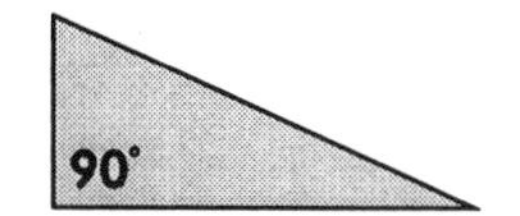

Now the amazing thing about these different triangles is that some are more than just one type.

Host: For example?

An isosceles triangle can also be a right triangle.

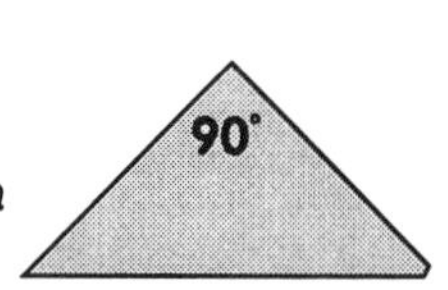

A right triangle can also be scalene.

Host: So THIS is where your multi-personalities come from. Tell us some other interesting properties.

Let me give you some problems and let you discover some properties for yourselves.

Host: I am game, and I'm sure our viewers are too.

You will need a ruler, paper and a pencil. Now, with the numbers 10, 7 and 5, and using your ruler, make a triangle whose sides are 10, 7 and 5.

Host: No problem. Here it is. What's so great about that?

Be patient. Now make one whose sides are 10, 5 and 3.

3 5

10

Host: I am sure that won't be a problem. 10, 5 and 3. Hmmm? The sides can't be joined.

You have just discovered one of the basic properties of a triangle. ***Any two of its sides must total more than the third side.*** *Otherwise a triangle can't be formed.*

Host: *Gee—* I never thought about it that way before.

Now let's learn something very special. In fact, this idea is over 5,000 years old. Make a triangle whose sides are 3, 4 and 5.

Now make one with sides 8, 15, and 17. What do these two triangles have in common, or should I say what type of triangle are they besides scalene? Study them carefully.

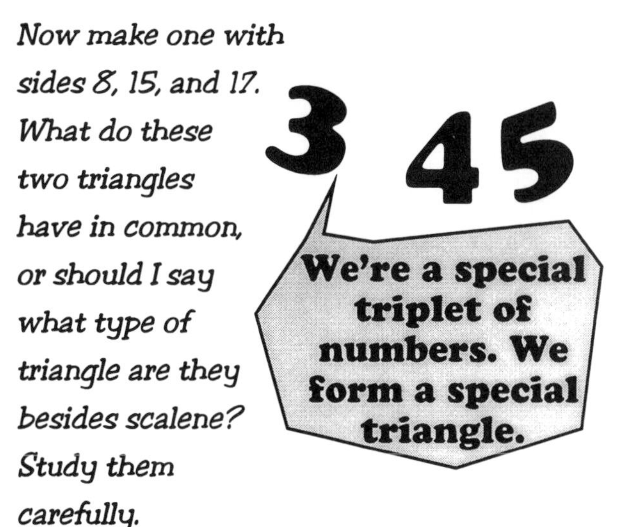

Right! They are right triangles. Certain triplets of numbers always produce right triangles. In fact, these triplets are so useful that they have their own name—

Pythagorean triplets. Double the triplet 3, 4 and 5 and make a triangle with 6, 8 and 10. Is it a right triangle? Try it with 5, 12, and 13. Try it by multiplying the triplet by any other number and see what shape triangle is formed? I'll let you explore that idea.

5 12 13

We are Pythagorean triplets!

Host: How are triplets used?

Well, if you know a certain set of numbers will always produce a right triangle, you now know a way to always make a right angle. And people from ancient times knew how important right angles were for building structures.

Host: Fascinating! Unfortunately our time is up. We need to have you back. Our phone lines are jammed with callers wanting to ask you questions. Thanks for joining us.

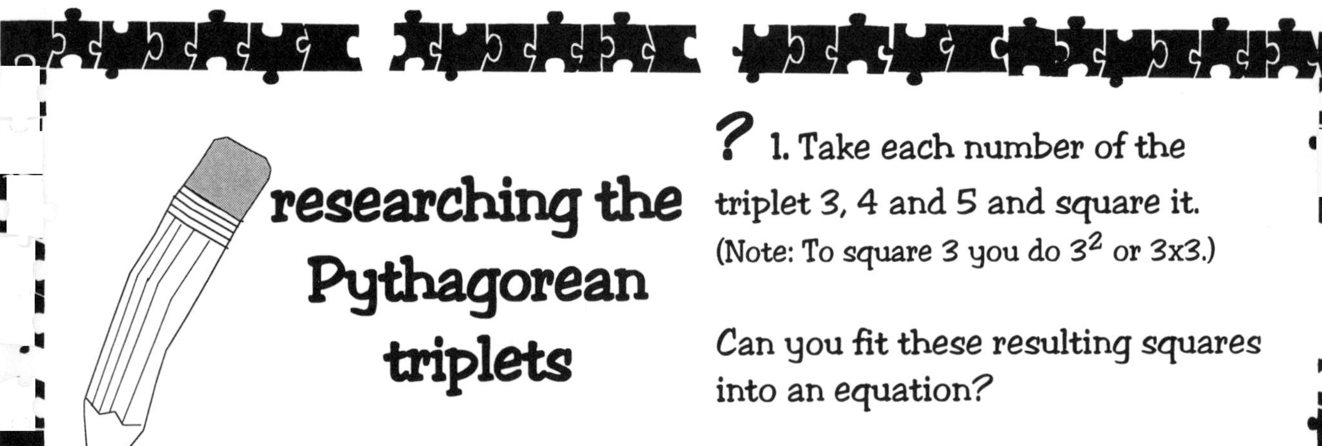

researching the Pythagorean triplets

? 1. Take each number of the triplet 3, 4 and 5 and square it. (Note: To square 3 you do 3^2 or 3x3.)

Can you fit these resulting squares into an equation?

? 2. Do the same with 5, 12, and 13. What equation do you get?

The equation which you found is called the Pythagorean equation. It is a way of testing whether the sides of a triangle will form a right angle. If the squared numbers of three sides of a triangle do not make an equation, then the triangle cannot be a right triangle.

? 3. Do three sides of lengths 4, 5 and 6 make a right triangle? Do the square numbers of these three sides make an equation?

Answers are given in the solutions sections at the back of the book

The subset party

It was the annual whole numbers reunion. This year the whole numbers had invited all sets that were subsets of **it** to come to the grand gala. The whole numbers insisted that any attending set list all ITS numbers in braces. The whole numbers, nicely listed in set notation, {0,1,2,3,4,5,6,7,8,9,10,11,12,13,14,15,...}, were observing all its subsets socializing. The even numbers led by 2, began line dancing subset {2,4,6,8,10,12,14,16,18,20,22,24,...}. Each element was very proud to be divisible by 2. On the other hand, the odd numbers were up to some of their old tricks. They were showing off how any two odd numbers could be added together and result in an even number. Besides other infinite subsets like the prime numbers and the multiples of 5, many finite subsets were also present. Among these subsets were the factors of

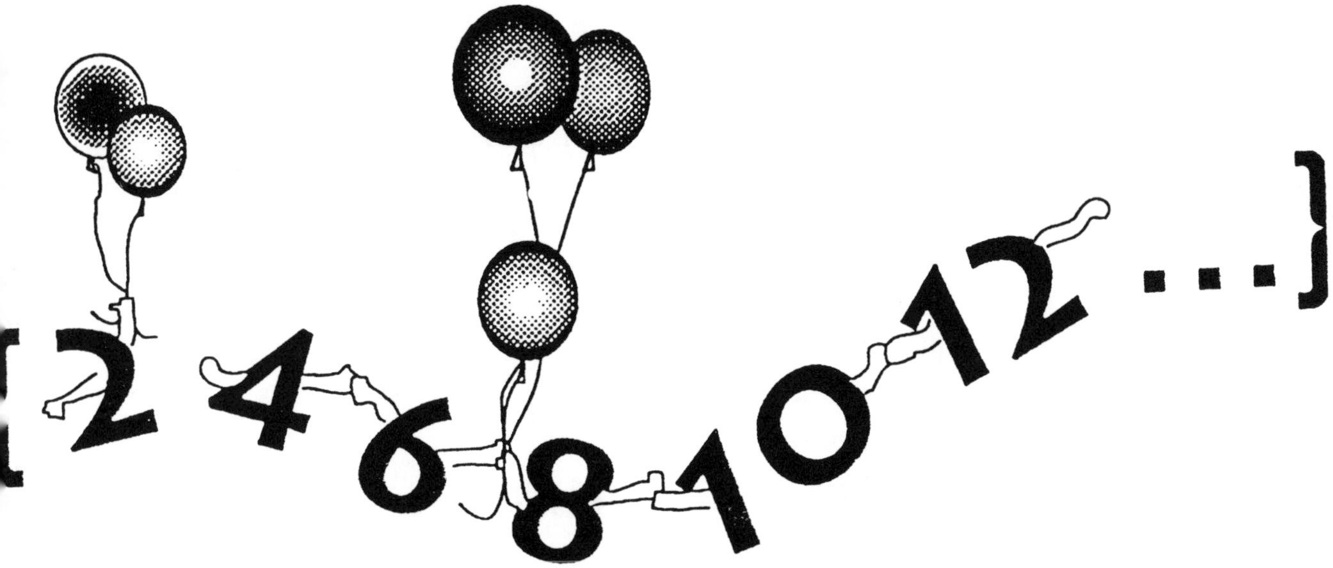

6, {1,2,3,6} and the set of whole numbers less than 8, {0,1,2,3,4,5,6,7}.

All these sets were having a great time laughing, singing and sharing stories of what new problems they were being used for since their last reunion. The perfect square numbers – {1,4,9,16,25,36,49,64,81,100,121,...} – were describing all the squares for which they were areas. The perfect cube numbers – {1,8,27,64,125,...} – were saying how cubes relied on them to figure their volumes. Even the set with zero – {0} – was sharing

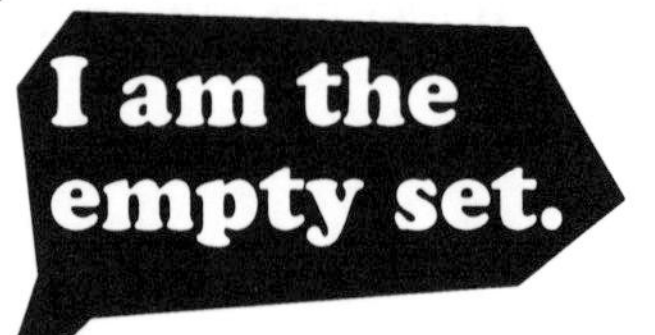

{ }

the types of problems it solves, "Imagine, I'm the only number that can solve these equations, ? +3=3 and nxn=0," {0} said.

Each set liked to flaunt its elements and explain how they were solutions to problems. All these sets were so proud to be subsets of the whole numbers.

Suddenly there was a commotion. Sets were whispering and pointing to something unusual. They surrounded the intruder. The even numbers were the first to speak. "Who are you and what are you doing here?" they asked, directing the question to a set that looked empty – { }. "I am here because I am a subset of the whole numbers," { } replied. "But there is nothing in your braces," the odd numbers pointed out. "True, because I am the **empty set**," { } said.

"What use is an empty set? What kind of problem do you solve?" the multiples of 3 set asked with a snicker.

"Like no problems," the set with the factors of 6 said laughing.

What use is an empty set?

{3}

"What type of problems have no answers," the set {1} taunted.

"I solve ***very special*** problems. You do not come across the problems I solve very often," { } continued.

"Give us an example of such a problem," demanded the set of whole numbers loudly in an effort to calm things.

"Well, all right. What is your solution to this equation: n+5=n?" { } asked.

All the sets patiently waited for someone to step forward and be the answer. But no set stepped forward.

"Well?" asked { }.

"There is no number to which you can add 5 and get the same number," replied the set {0} in a perplexed tone. "Try any of the numbers in the set of whole numbers, none will work. Even non-whole numbers won't work. There is no fraction or any number," {0} explained. All the sets looked at one another. "That's true," they all said, astonished.

After a long silence the whole number set said, "The **empty set** is the answer to this problem. The **empty set** is one of us," it continued to explain. "You are welcome here at our reunion, **empty set**."

"May I add one comment?" { } asked the whole numbers. "Of course," the whole numbers replied.

"Because I am so special, mathematicians have given me an additional symbol name. It is this symbol, ϕ. And further, I happen to be a subset of every set. So I can attend any set party."

Because I am so special, mathematicians have given me an additional symbol name.

"Of course," the factors of 6 —{1,2,3,6} —said. "You're absolutely right," with a pleasant tone to its voice.

With that comment sets gathered around { } and began asking it questions and offering it refreshments. They seemed happy to have { } among them, especially now that they understood it.

The walk of the seven bridges

The year was 1735.
The place was the town Königsberg, situated along the Pregel River. In fact, the river ran right through Königsberg, so the different parts of the city were connected by bridges. There was the Shopkeepers bridge which led to the street of shops of Königsberg. The blacksmith had a bridge which led to his place. The Honey bridge connected Kneiphof Island to the other island. The other four bridges were the Wooden bridge, the Green bridge, the High bridge and the "Guts" Giblet bridge. Seven bridges in all, joining the two islands of Königsberg with the rest of the town. Königsberg was a charming town, with its islands and bridges. Most people loved to take walks along the river and the islands. Indeed, it had become a Sunday tradition to take the walk of the seven bridges. *The walk of the seven bridges* was no ordinary walk, but a walk with a specific problem in mind—*to discover a*

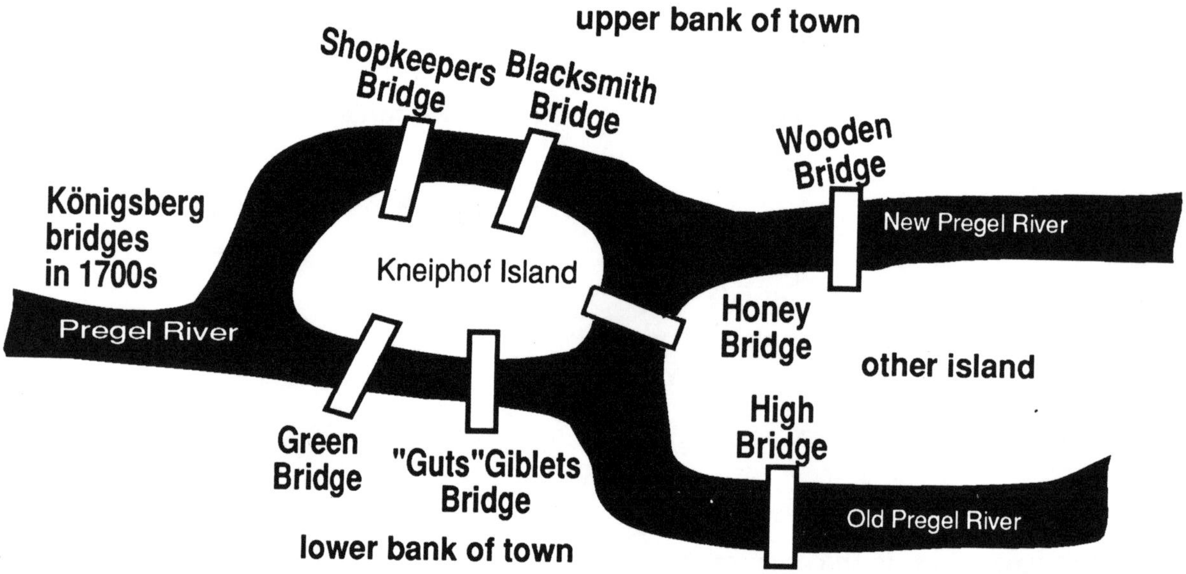

path that would cross all seven bridges without recrossing any bridge. The blacksmith had tried a number of times to find a path, but he was not successful. The cobbler, the baker's son, the seamstress and her family– all had tried, and all had failed. Even though the townspeople had been taking the walk of the bridges for years, no one had solved the problem. As a result, the Königsberg bridge problem had become quite well known throughout Europe. It seemed to the townspeople that no one was going to solve their problem. When the problem came to the attention of Swiss mathematician Leohnard Euler, it intrigued him. In 1735, he presented his solution to the Russian Academy. His work on this problem set other ideas into motion and launched a new field of mathematics called topology.

How did he solve the problem? He decided not to bother trying to walk out the solution, but instead to draw a diagram of the problem. He studied the map and the bridges, and finally came up with this simple diagram. The seven arcs represented the seven bridges. The four dots represented the parts of the town that the bridges connected. He figured if he could trace over his diagram with a pen without ever doubling back, he would have discovered the path.

Try it out. What is your solution?

See the back of the book for Euler's solution.

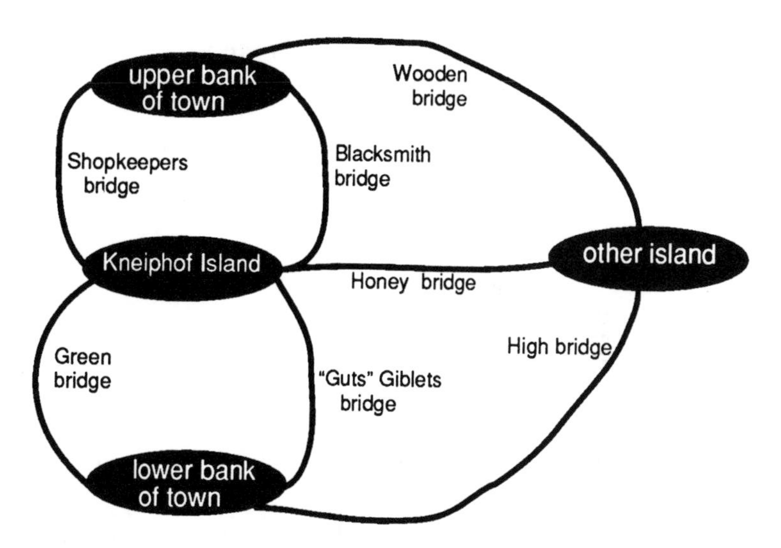

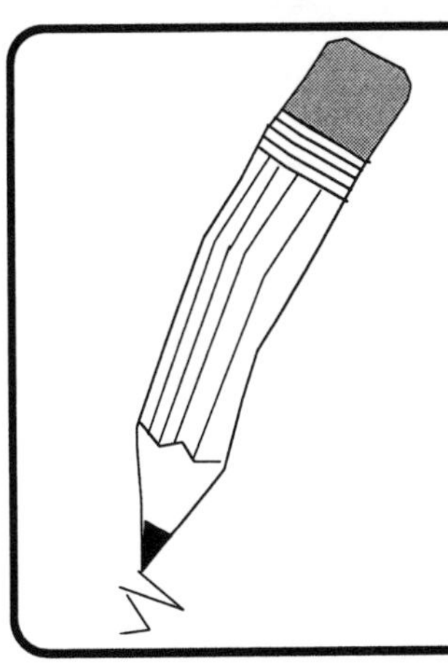

researching:

Look in your local library or on the internet for additional information on the Könisgberg bridge problem and mathematician Leonhard Euler.

The tri-hexa flexagon

There are many objects in mathematics that seem almost magical.

The tri-hexa-flexagon is one of them. With a strip of paper and some fancy folding you can create an object that has three different sides. It gets its name as follows: *tri* for the three sides, *hexa* for the six edges of the hexagon formed, and *flexagon* because you reveal the sides by flexing or folding the paper. The tri-hexa-flexagon is one of many flexagons. If you enjoy making this one, why not see if your library has a book on flexagons. Discover some of the other types of flexagons.

You can use them to write messages, as greeting cards, stationery, or just for fun.

Make a copy of the top strip and label its equilateral triangles on the front and back sides as shown.

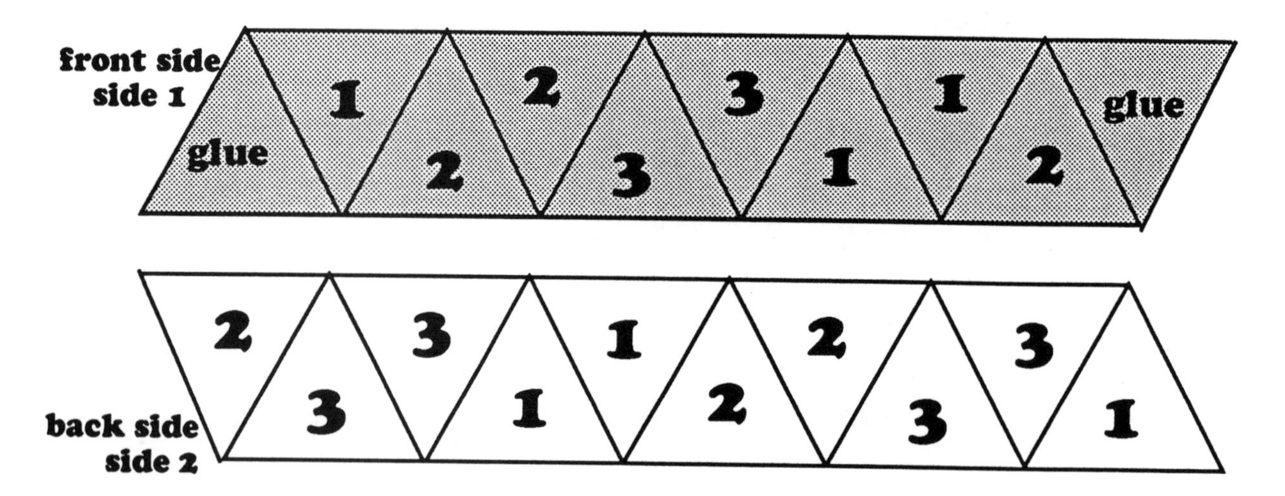

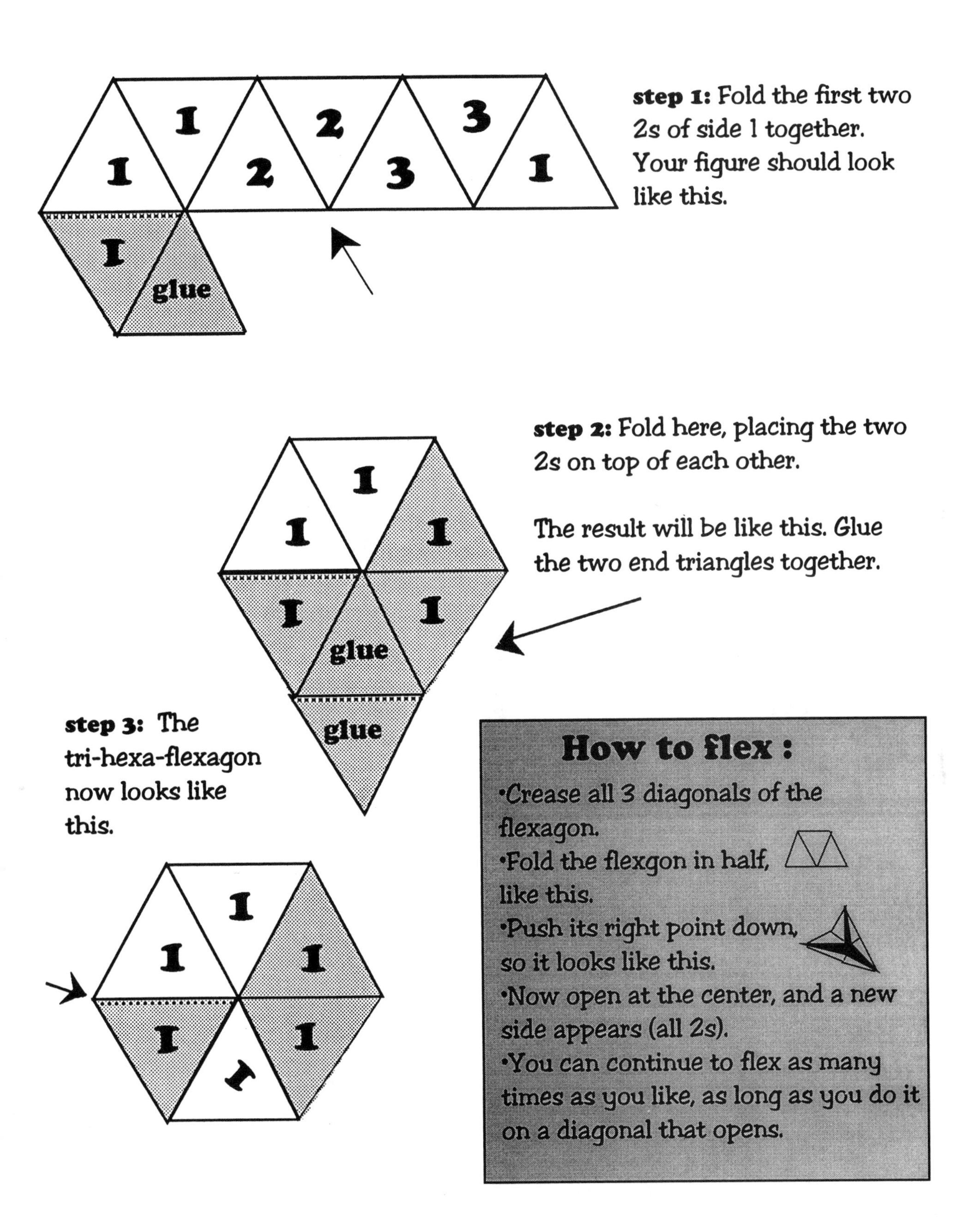
1
1
2
2
3
3
1
1
glue
step 1: Fold the first two 2s of side 1 together. Your figure should look like this.
step 2: Fold here, placing the two 2s on top of each other.
The result will be like this. Glue the two end triangles together.
1
1
1
1
1
glue
glue
step 3: The tri-hexa-flexagon now looks like this.
1
1
1
1
1
1
How to flex :
•Crease all 3 diagonals of the flexagon.
•Fold the flexgon in half, like this.
•Push its right point down, so it looks like this.
•Now open at the center, and a new side appears (all 2s).
•You can continue to flex as many times as you like, as long as you do it on a diagonal that opens.

Learning about the geometry of nature

Look at this fern. It certainly does not look like a square, a rectangle or a circle. This fern cannot be drawn even if triangles or trapezoids are used. Believe it or not, this fern was made from a mathematical object called a fractal.

What is a fractal? A fractal is a shape that continually duplicates itself over and over but in smaller and smaller sizes. It never stops generating itself. It does it infinitely many times, following a rule or a mathematical equation. In fact, when you see a fractal generated on a computer monitor, it seems to be growing because it is. As we grow our body adds cells. When fractals grow they add ever smaller fractals.

Fractals were first discovered in the late 1800s. At that time many mathematicians called them mathematical monsters because fractals had such unusual properties. Today fractals are being used in many areas such as biology, environment, medicine. They can be used to describe things that are constantly changing or growing, such as things in nature. Fractals have come to be called the *geometry of nature.*

Let's find out how a fractal was used to make a tree. This is the starting shape that was used.

This shape was continually reproduced in ever smaller sizes on each branch of the tree.

Here are the first three stages of this fractal tree.

Using a crayon or highlighter see if you can find these different stages in the tree. Can you see even smaller sizes of the fractal in the tree? Get a magnifying glass and look at a limb of the tree. See how that shapes repeats itself over and over again?

The fractals we see here are called geometric fractals. There are also random fractals made using other mathematical methods. Random fractals like geometric fractals never stop growing or changing. Unlike geometric fractals where the same process is applied over and over, random fractals are formed by changing a step or form randomly.

more about fractals

Many computer programs have been designed to generate fractals.

They are very exciting to work with because the computer let's you see the fractals grow before your eyes.

You may find some free information and software on the internet.

Be sure not to miss seeing a fractal evolve before your eyes. The fractals shown on the previous pages were generated using a computer. Here are some others that have been made using the computer.

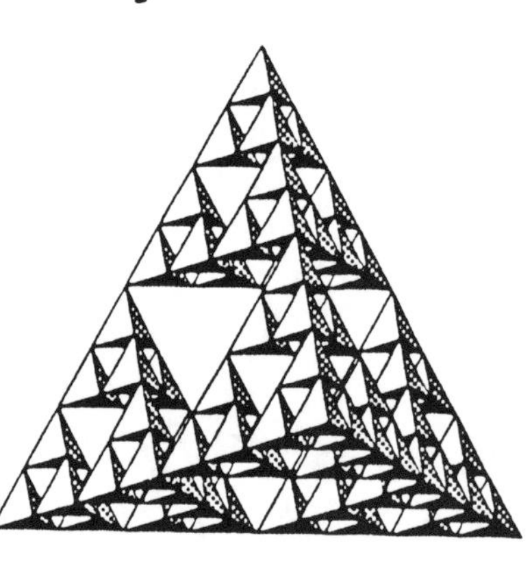

researching:

•Go to the library a check out a book on fractals and look at examples of fractals.

•Use the library computer or your own to find fractals programs available on the internet.

Discovering the secret of the diagonals

Diagonals have always been shifty characters. They never do anything straight or give you a straight answer. They have an angle to everything, since they are always at an angle. So when pentagon wanted to find out how to tell ahead of time how many diagonals it was going to have, the diagonals were no help at all. "There is no way to tell without just drawing us in. There is no simple way because we are very very complex special objects," they declared in very snobbish voices.

Triangle, what can you tell me about diagonals?

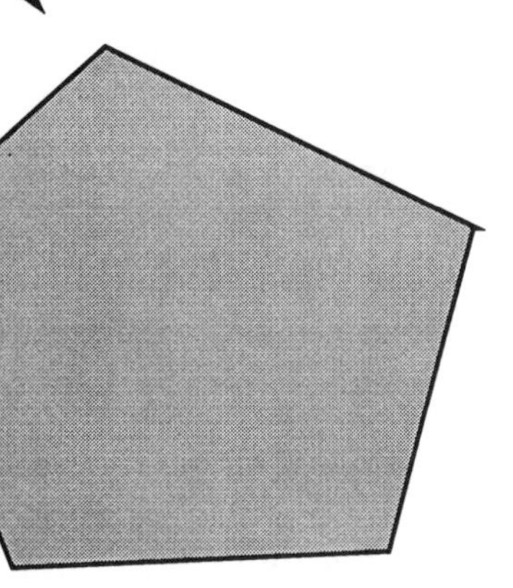

Pentagon, being the most psychic of the polygons, sensed there had to be a shortcut. So what did Pentagon do? It decided to talk to other polygons who already had their diagonals drawn in, and see if it could get some help. First Pentagon approached the smallest-sided polygon, Triangle. "Triangle, what can you tell me about diagonals?" "All I can say is that I am glad I have had nothing to do with diagonals, since I don't have any – as you see none can be drawn in me," Triangle replied. "I see what you mean," Pentagon noticed, and decided to go to see Quadrilateral, the four sided polygon. "Quadrilateral, what do you have to say about diagonals?" Pentagon

All I can say is that I am glad I have had nothing to do with diagonals.

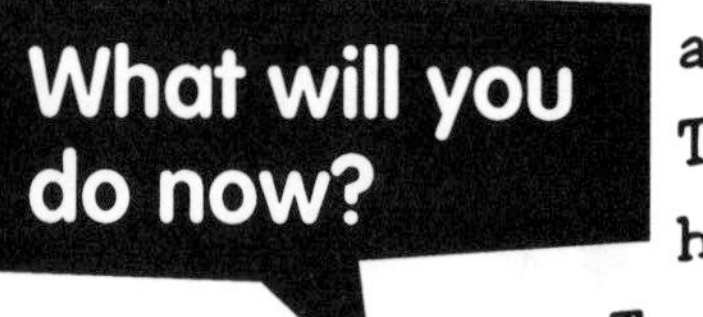

asked. "Two. Two is a all I have to say. Two is what I have always had. Two diagonals, no more no less, that's all I know. If you want more information, speak to Hexagon. I am sure it can help you in your quest." With that Quadrilateral turned abruptly and left, leaving Pentagon standing alone.

Not easily discouraged, Pentagon next went and knocked on Hexagon's door for help. Hexagon opened the door, and cordially invited Pentagon in for a chat. "Pentagon, I understand you are interested in finding the secret that the diagonals have. I do hope you discover their secret. Every time someone wants to know how many diagonals I have, they have to draw them in on me. And frankly, I wish they could figure out the answer without burdening me with 9 diagonals. What will you do now?" Hexagon asked Pentagon. "I guess I'll have to ask for some help from our readers. Perhaps they can find the secret of the diagonals by filling in the missing information, and thereby discover the formula of the diagonals."

I guess I'll have to ask for some help from our readers.

1) Draw in the different diagonals for the polygons pictured.

2) Fill in the chart.

3) Guess the equations for the hexagon and heptagon that go in the gray column.

4) Make a rule for finding the number of diagonals without drawing them in.

5) Use your rule to find the number of diagonals for an octagon and a nonagon. Does your rule work?

A solution is given at the end of the book.

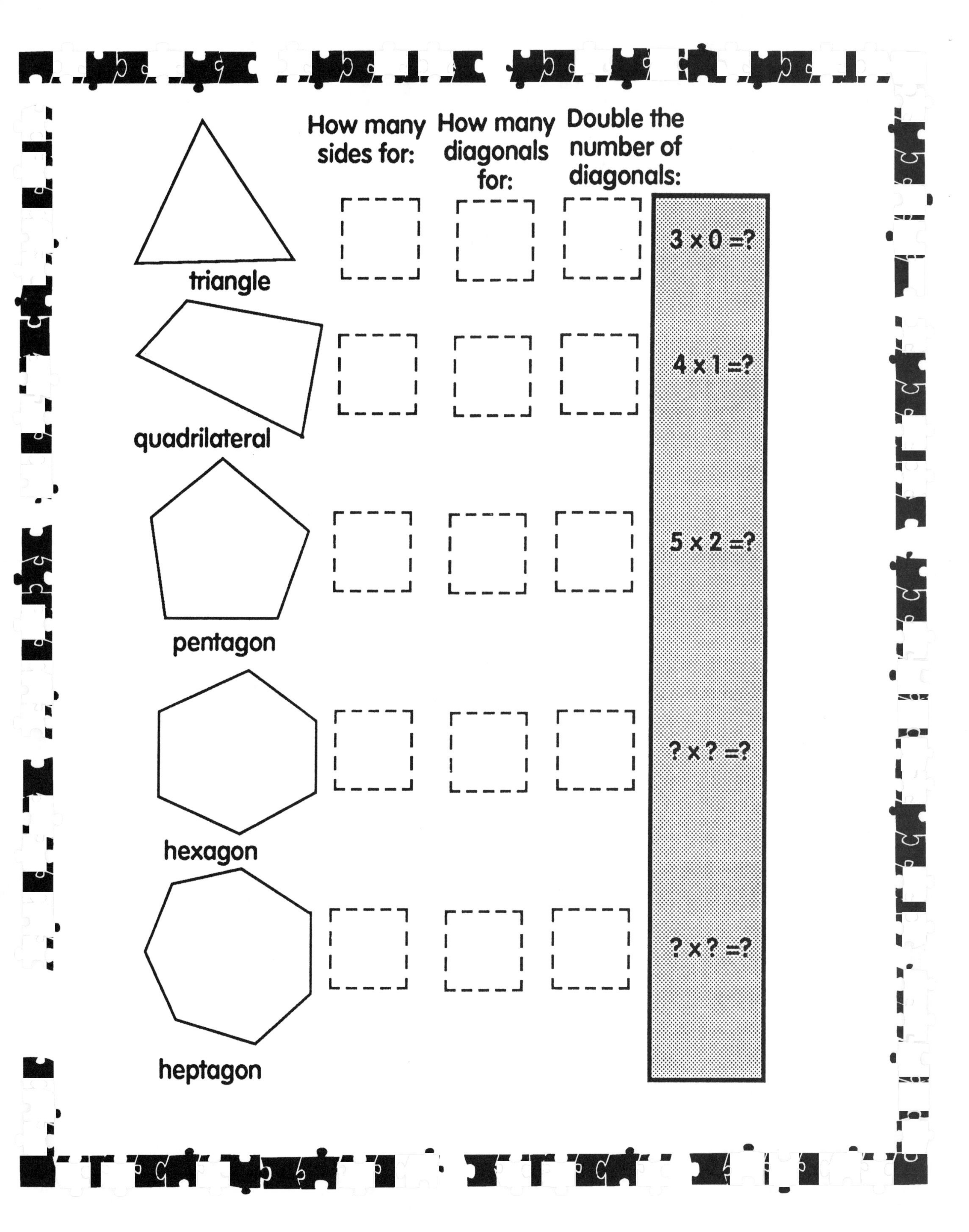
How many sides for:
How many diagonals for:
Double the number of diagonals:
triangle
3 x 0 =?
quadrilateral
4 x 1 =?
pentagon
5 x 2 =?
hexagon
? x ? =?
heptagon
? x ? =?

Two dimensions change to three – plus other optical illusions

It is amazing to become aware of the tricks our eyes can play on us. There is an old saying, "When I see it, I'll believe it." But often things we perceive with our eyes are not actually how they exist. For example, looking at these two figures, it appears that the one below is larger. But to be sure, trace over it with a piece of paper and see if it is larger than the upper one. You will find they are exactly the same size.

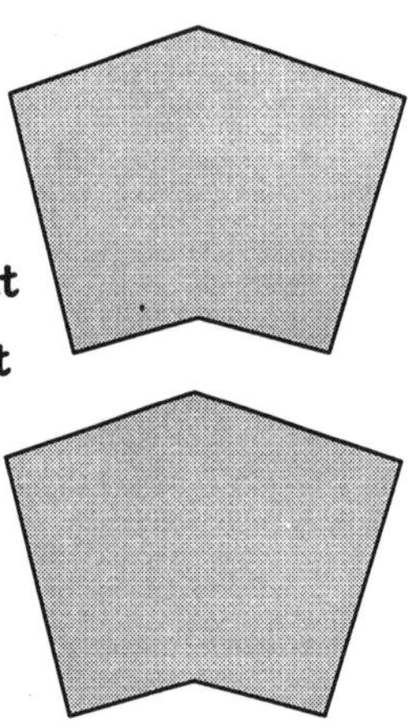

There are many optical illusions that have been observed or created over the years. Let's look at a few.

Make a rhombus. Using it as a template draw the rhombus together in different positions, all fitting together in various ways.

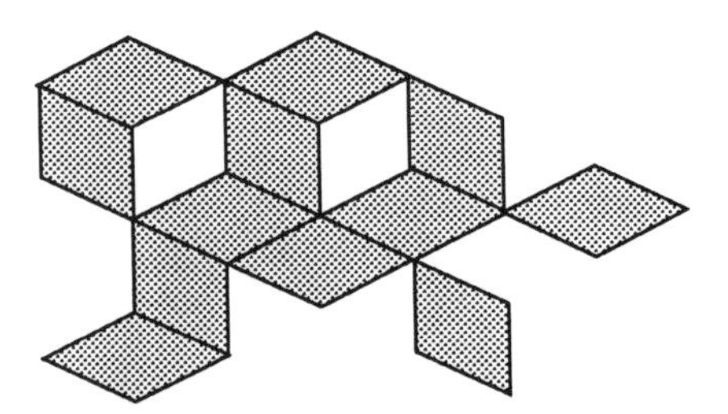

Before you know it you will have created an optical illusion that fluctuates between being flat (2-dimensional) and poking out (being 3-dimensional). After staring at it for awhile your mind will become tired of viewing it one way and flip to viewing it another way. (If you have access to a computer, try designing the rhombus in a drawing program and rotate it and fit it together on the computer's monitor.)

Here is another type of moving optical illusion. It is called an *afterimage.* Stare at the black rectangle for about 30 seconds. Then look at a dark wall or surface ahead of you for about 10

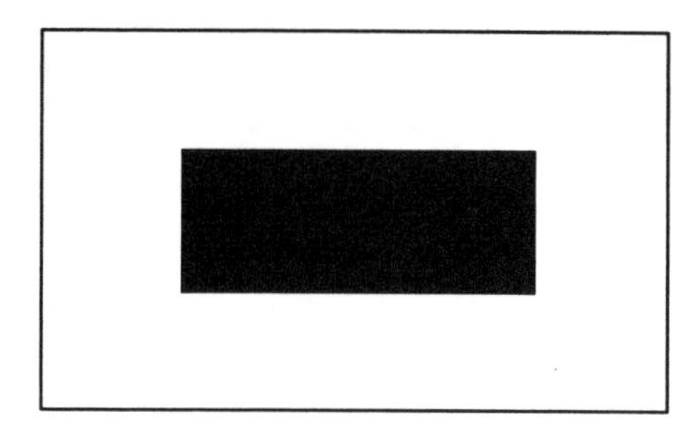

seconds. What happens?

Right! A rectangle appears on the dark surface, but it is no longer a black rectangle, but appears as light or white. Pretty amazing? We cannot always trust our eyes because our mind may play tricks on us. It is also important to use our logic and actual measurements when we are confronted with different optical phenomena.

Here's how to make a moving illusion.

STEP 1 -Take a sheet of paper which has ruled lines about 3/8 of an inch apart. Cut out every other strip. The sample template shows the remaining sheet of paper in gray. The white strip at the bottom represents a strip of paper that is glued to keep the strips from flapping around. Using this template, you can create a moving illusion by carefully following steps 2 through 4.

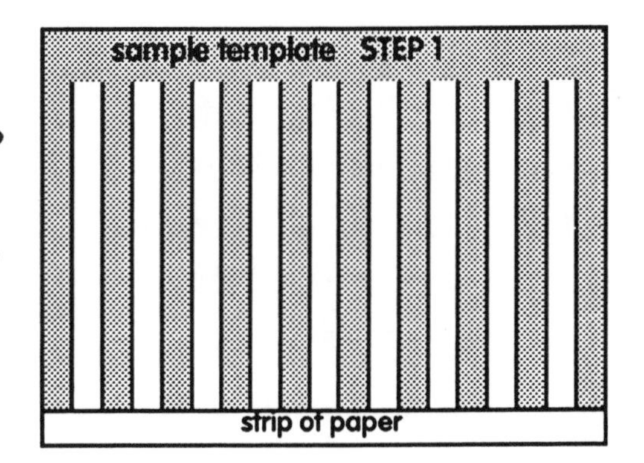

STEP 2 -You draw your object in two stages. Place the template on a blank sheet of paper and draw a bird as shown. Note that nothing is drawn where the template strips are.

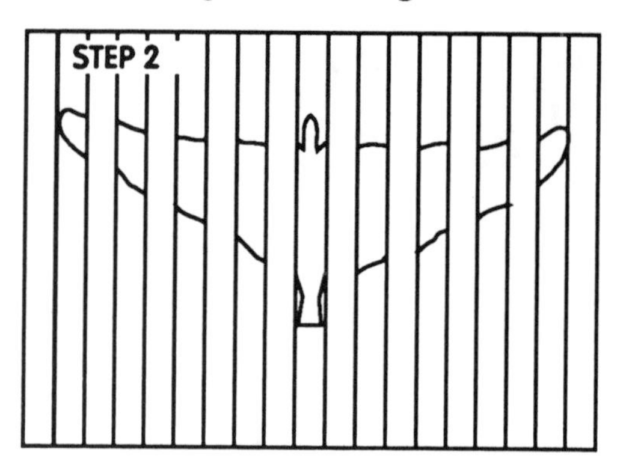

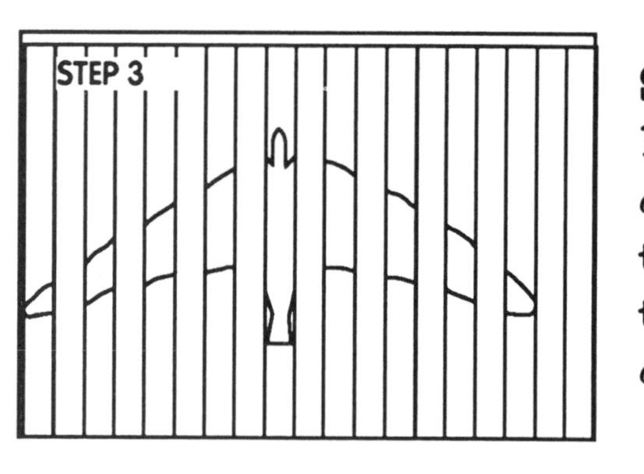

STEP 3- Slide your template 1/2 inch over, so your drawing is concealed by the template's strips. Now draw the bird with its wings down.

STEP 4- After doing steps 2 & 3, here is how your drawing would look without the template on top of it.

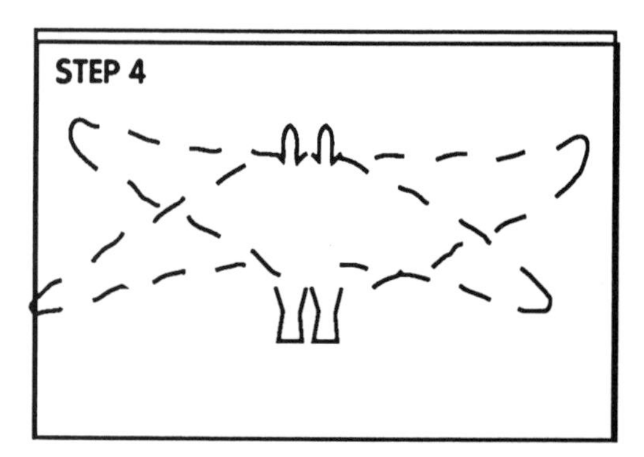

Now to see the bird in motion, ***slowly*** slide the template across the picture while staring at the paper. Slide it back and forth. The bird's wings will appear to flap.

In the same way, using the same template and a fresh sheet of paper, draw your own moving illusion. Perhaps it will be a person walking, a ball bouncing, or anything you think of.

WHAT'S A ZILLION? A look at really big numbers

How much is a zillion?

How about a *few, some, a lot, many?* What size are these numbers. What numbers do these names stand for? None of them are numbers. Some of us

a zillion=????

get confused and think a zillion must be a number since it sounds so much like a million. People began using the word zillion to express a really large amount, but it is not a specific number. Let us look at the big guys—the really large numbers which have names.

The others— *zillion, many, a lot, some, few*—are not a specific amounts. They are indefinite amounts. Indefinite means not defined, not specified.

So the next time you hear someone say a zillion, you might want to ask them to write out the number for you. You know they won't be able to. So you can tell them about some of the real number names that end in —illion and write out their number forms; that is if your hand doesn't get too tired.

the big numbers

hundred	100
thousand	1000
million	1,000,000
billion	1,000,000,000
trillion	1,000,000,000,000
quadrillion	1,000,000,000,000,000
quintillion	1,000,000,000,000,000,000
sextillion	1,000,000,000,000,000,000,000
septillion	1,000,000,000,000,000,000,000,000
octillion	1,000,000,000,000,000,000,000,000,000
nonillion	1,000,000,000,000,000,000,000,000,000,000
decillion	1,000,000,000,000,000,000,000,000,000,000,000
undecillion	1,000,000,000,000,000,000,000,000,000,000,000,000
duodecillion	1,000,000,000,000,000,000,000,000,000,000,000,000,000
tredecillion	1,000,000,000,000,000,000,000,000,000,000,000,000,000,000
quattuordecillion	1,000,000,000,000,000,000,000,000,000,000,000,000,000,000,000
quidecillion	1,000,000,000,000,000,000,000,000,000,000,000,000,000,000,000,000
sexdecillion	1,000,000,000,000,000,000,000,000,000,000,000,000,000,000,000,000,000
septendecillion	1,000,000,000,000,000,000,000,000,000,000,000,000,000,000,000,000,000,000
octadecillion	1,000,000,000,000,000,000,000,000,000,000,000,000,000,000,000,000,000,000,000
novemdecillion	1,000,000,000,000,000,000,000,000,000,000,000,000,000,000,000,000,000,000,000,000
vigintillion	1,000,000,000,000,000,000,000,000,000,000,000,000,000,000,000,000,000,000,000,000,

How would this number be read?

24,700,631,500,070,002,100,344,298,023,185

The answer is at the back of the book?

MATH PUZZLES, GAMES, and TRICKS

Bagels are not just for eating – the game of bagels, fermi, pico

If you think bagels are a delicious kind of bread often spread with cream cheese, then try this logic game with a friend.

Bagels, Fermi, Pico sounds like someone asking if their friends Fermi and Pico would like some bagels. But to the contrary– it is a great logic game that does not require any special playing board, pieces or electronic equipment. You just need a friend, two strips of paper and two pencils.

HERE IS HOW IT IS PLAYED.

1) Each of you picks a 3-digit number and writes it secretly on a strip of paper. Make two folds to conceal the number you chose.

Object of the game: The winner is the first to guess his or her friend's number.

As each guess is written, one of the following clues of *Bagels, Fermi* or *Pico* is written down next to the guess.

Bagels means that no digits of the guess given is correct.

Fermi means one digit of the guess is correct and its location is also correct.

Pico means a digit is correct, but its location is wrong.

For example, suppose the secret number picked was *372*, and you guessed the number *275*. Your friend would write the

Here is a sample game of Bagels, Fermi, Pico

The secret number 372

372

???

guesses	clues	
275	pico fermi	This clue tells me that I have guessed two digits and that one of them is in the correct place.
175	fermi	I decide to see if the 7 and the 5 are the two right digits. Now this clue tells me that either the 7 or the 5 is correct.
465	bagels	I decide it may be the 5, and make this guess. Now the clue tells me none of these digits is correct. So the 7 was right and the 2 was in the wrong place.
972	fermi fermi	So I relocate the 2, keep the 7 where it was and choose 9 for the hundreds place. But the clue tells me that 9 is the wrong choice.
872	fermi fermi	So now I decide to try 8. But again the clue tells me that the 8 is wrong.
372	fermi fermi fermi I GOT IT!	3 and 0 are the only digits left, but a whole number cannot begin with 0, so 3 is my choice. I finally guessed it.

clue Pico Fermi or Fermi Pico. The order of the clue words is not important.

Keep track of your guesses and responses on your strip of paper. Using logic and clever guesses, the player finding the opponent's secret number with the least number of guesses is the winner.

If you like, you can start your first game by picking two-digit secret numbers. Make games harder and harder by increasing the number of digits. Develop some interesting strategies for your guesses. For example, if your guess was 438 and the response was bagels, right away you know none of the digits is right, so you'll never use a 4 , a 3 or an 8 in your guess again.

Pencil tricks & mathematics

Topology is a field of mathematics that has many unusual objects. In topology a circle can become a square. In topology we can transform a strip of paper into a single sided object. These two pencil tricks rely on topology to work their magic. Try them out, and astonish your friends.

The button hole pencil problem

What you need:

- a pencil
- tape
- a piece of string less than twice the length of the pencil
- a sweater, blouse or shirt with button holes

What to do:

First thread the string through a button hole, as shown.

Attach the pencil to the ends of the string with tape, as shown.

Problem: Remove the pencil from the button hole without cutting the string.

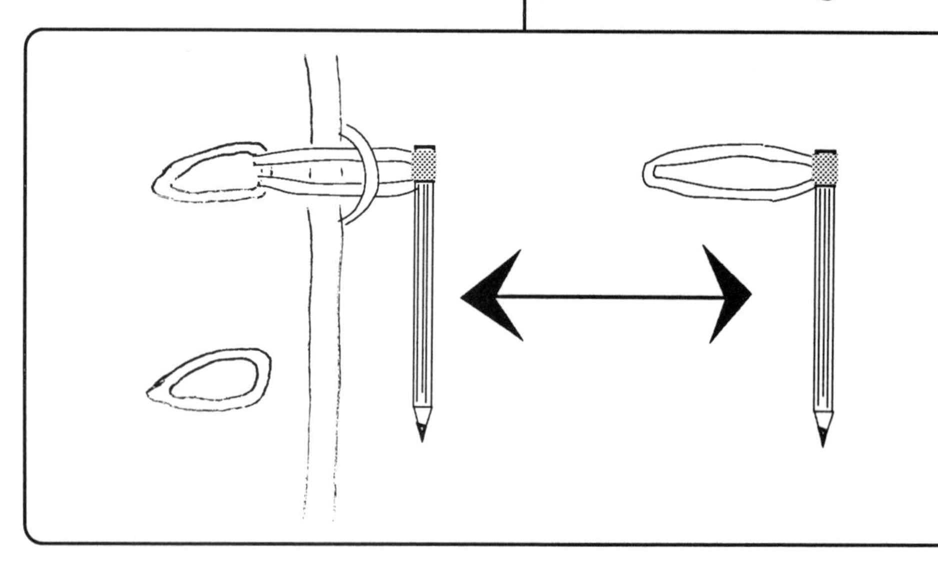

The solution is given at the back of the book.

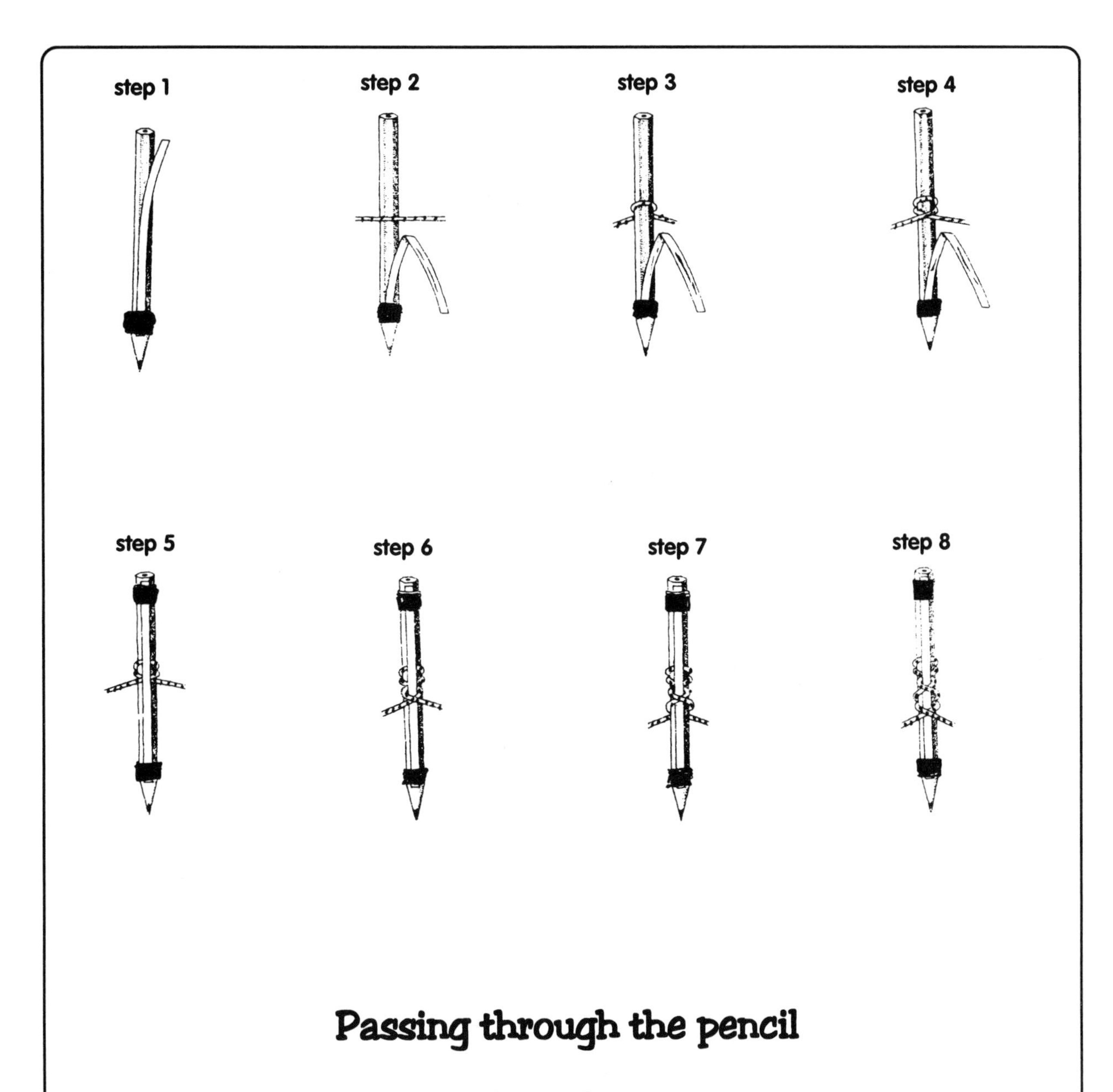

Passing through the pencil

What you need:

- a pencil
- string
- a thin strip of paper
- tape

What to do:

- Attach the strip of paper to the pencil with a piece of tape, as shown.
- Loop the string around the pencil and strip, following the step illustrated.
- Now tug on the string and it should pass magically through the pencil and cut the strip of paper.

The binary cards

When you first try out these cards, you may think there is something magic about them, but their magic is all mathematical.

Here's how they work. Ask a friend to pick a number between 0 and 64, but not tell you the number. Then hand your friend the six cards, and ask him or her to look at these cards and give you back only the cards which have the number picked somewhere on them. Then place your special math card on top of the stack your friend gave you. Total the numbers that appear through the holes. That sum will equal the number picked.

WHAT'S THE SECRET?

Each card is made in a special way. In the story *Penrose meets the Os and 1s* you learned how numbers could be written in a number system that uses only the digits 0 and 1. You also learned that this number system is called base two. If you wrote the number 10010, to find its value in our base ten system, you would just convert it by using its place value.

_	_	1	0	0	1	0
64	32	16	8	4	2	1

so **10010** in base two is

16+2=**18** in base ten.

* * *

Look at each of the six cards on the next page. Circle the smallest number that appears on each card. Notice that these numbers are **1**, **2**, **4**, **8**, **16**, and **32**.

Each of these cards represents a place value in base two. Now pick a number between 0 and 64. Let's say you picked 50. Find which cards have 50 on them. 50 appears on the cards where the 2, 16 and 32 are circled. That's because 32+16+2=50. These cards automatically write the number picked in base two, 50 equals 110010 in base two which equals *one* **32** + *one* **16** + *zero* **8** + *zero* **4** + *one* **2** + *zero* **1**.

The holes on the magic math card appear where the **1, 2, 4, 8, 16,** and **32** are located on their cards. So when you have the cards which appear with the picked number, the magic math card gives you the place value for that number in base two, and the sum converts it to base ten.

Test some numbers out. Then try it on your friends.

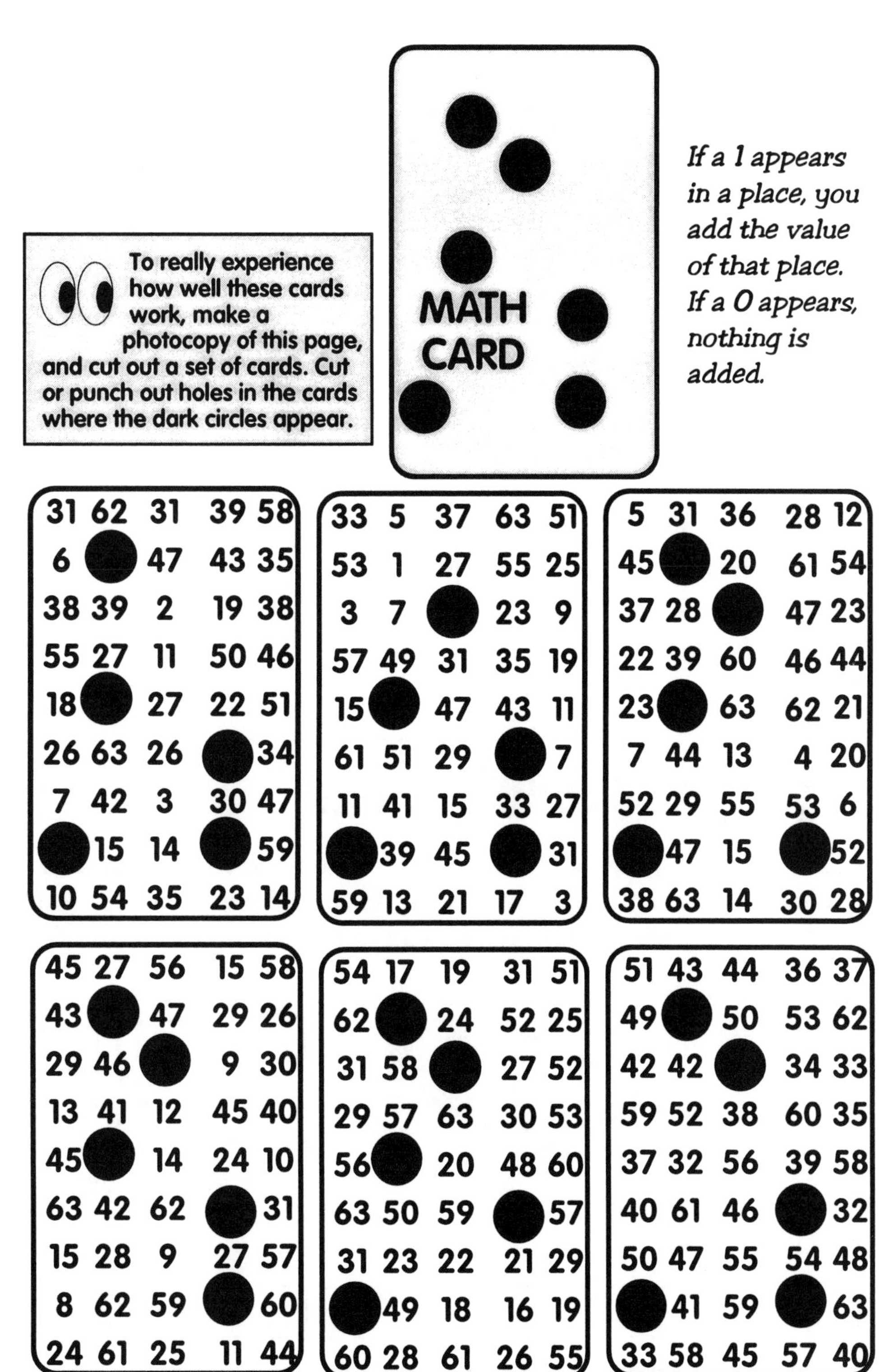

Brain exercises

These puzzles are meant to give your brain a workout and to have fun at the same time.

You can check your answers with those at the back of the book.

The magic triangle

In a magic triangle, the numbers on each side will total the same amount. Place the numbers 1, 2, 3, 4, 5, 6, 7, 8, and 9 along the three sides of the triangle so that each side totals 20.

Finding a path

You can only use gate 1 when coming or going to the brick house. Gate 2 is only used for the white house, and gate 3 for the gray house. Find a path from each house to its gate so that no path crosses another.

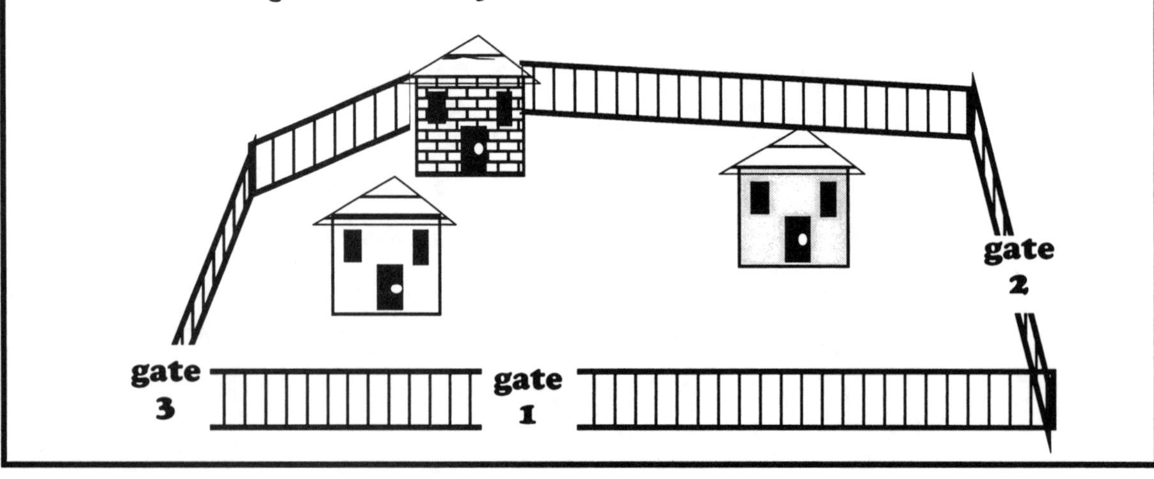

The Pail Problem

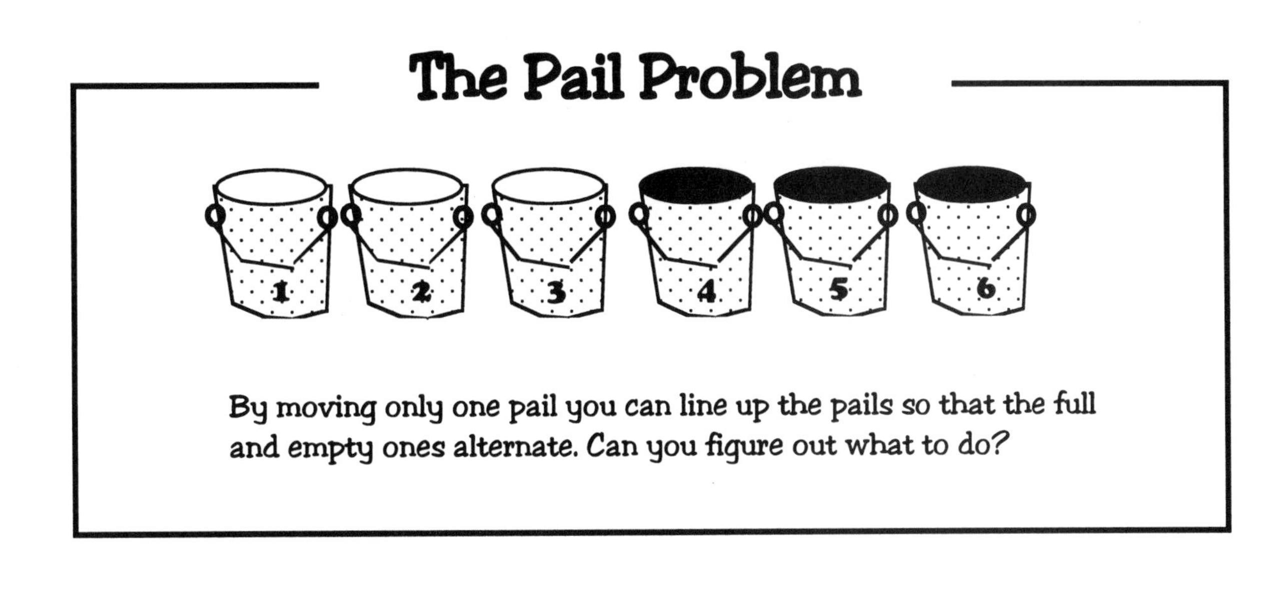

By moving only one pail you can line up the pails so that the full and empty ones alternate. Can you figure out what to do?

The line-up puzzle

Rules:

1. The black disks can only move to the right, and the gray can only move to the left.
2. A disk can move only one square at a time.
3. Only opposite colored disks can jump one another.

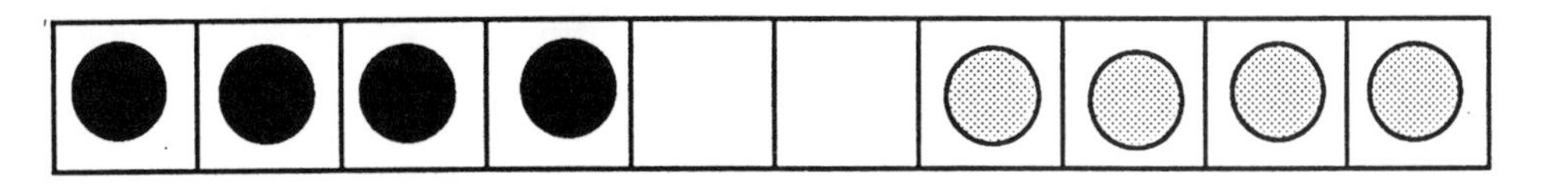

Object: Move the disks so that they are lined-up as shown below.

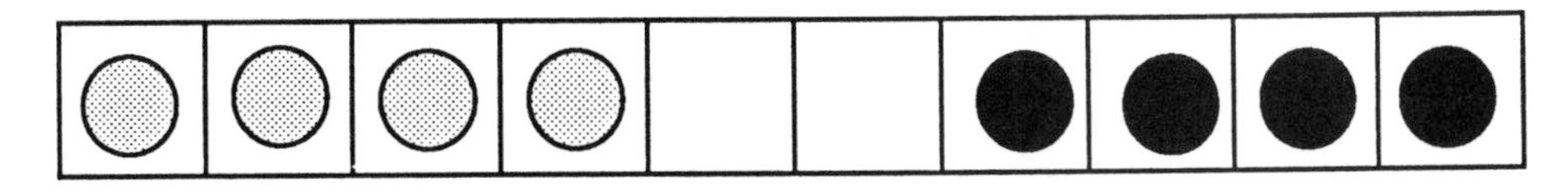

Hint: Try a smaller version of the puzzle first.

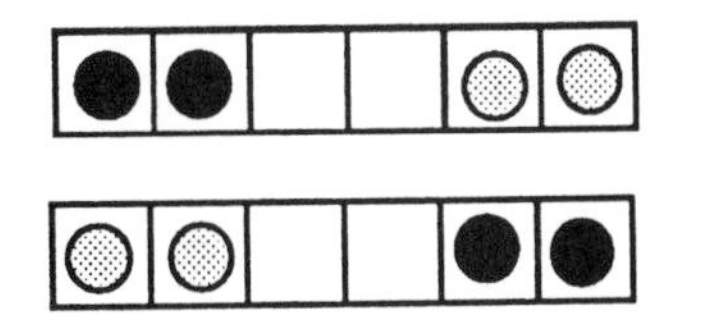

The game of sprouts

To play sprouts, all you and a friend need are two pencils and a sheet of paper.

The game begins with two or more dots, called spots.

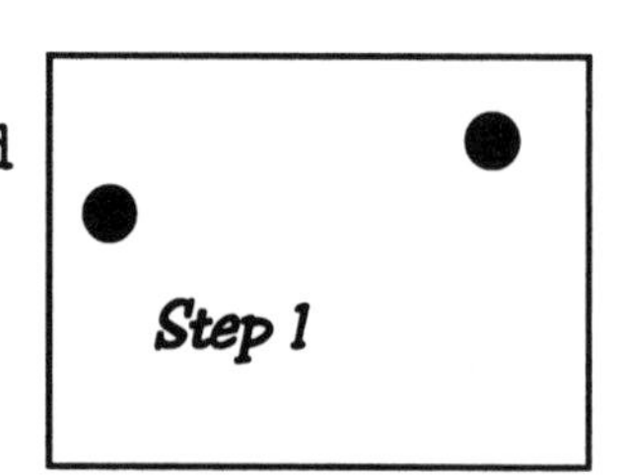

Step 1

Step 2: *Draw an arc. These are the two possible starting moves.*

Taking turns, each player draws an arc from one spot to another,

or from one spot to itself.

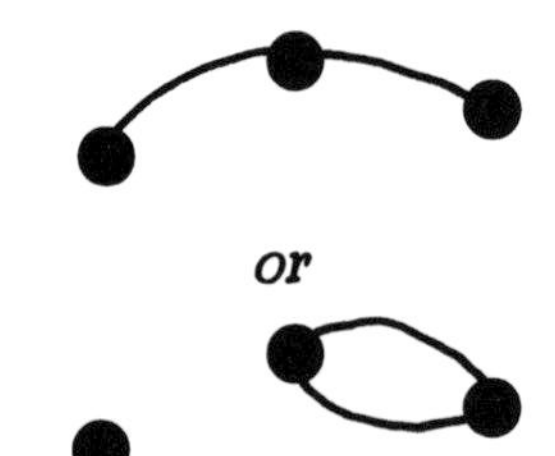

With each new arc a new spot is drawn on the arc by the player who drew that arc.

Step 3: *Draw a spot on the arc you just drew.*

RULES:

(1) No arc may cross itself or pass through another arc or spot.

(2) No spot may have more than 3 arcs coming from it.

When it does, it is circled and can no longer be used. It is called a dead spot.

The player unable to draw an arc is the loser.

You can make the game longer by adding more starting dots or by changing rule 2 by making a spot dead when 4 arcs come from it.

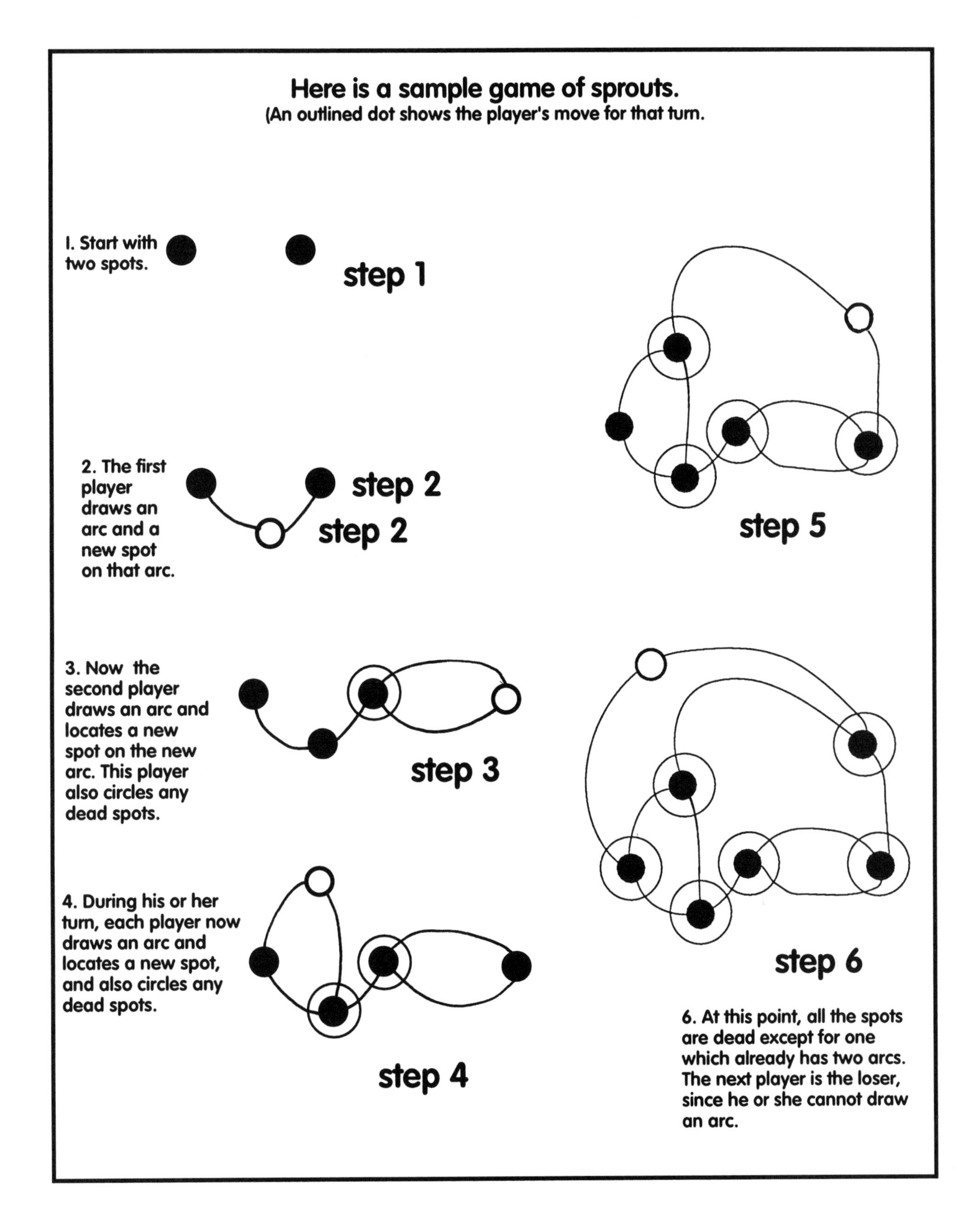
Here is a sample game of sprouts.
(An outlined dot shows the player's move for that turn.
I. Start with two spots.
step 1
2. The first player draws an arc and a new spot on that arc.
step 2
step 2
3. Now the second player draws an arc and locates a new spot on the new arc. This player also circles any dead spots.
step 3
4. During his or her turn, each player now draws an arc and locates a new spot, and also circles any dead spots.
step 4
step 5
step 6
6. At this point, all the spots are dead except for one which already has two arcs. The next player is the loser, since he or she cannot draw an arc.

What makes 9 so special?

Plus some tantalizing number tricks.

What's so special about the number nine?

There are many fascinating things that happen when you work with 9s. Study the different number patterns that form when nines are used.

Here are some more amazing properties of 9.

Writing out the product 9 times each of the digits from 0 through 9 is very tricky:

9 x 1 = 09
9 x 2 = 18
9 x 3 = 27
9 x 4 = 36
9 x 5 = 45
9 x 6 = 54
9 x 7 = 63
9 x 8 = 72
9 x 9 = 81
9 x 10 = 90

Just write out the digits 0 through 9 in the first column. Next to these now write the digits in the next column, but in reverse order.

ALSO NOTICE, the sum of the two digits of the product always totals_?_.

What else can nine do?

Take any three digit number whose ones and hundreds place digits are different. For example, 285. Reversing the digits, we get 582. Find the difference between the two numbers. 582-285=297. You ask, what's so great about that? The middle digit will **always** be 9, and the other two digits will **always total** 9.

Number patterns with 9s

999,999 x 2 = 1,999,998
999,999 x 3 = 2,999,997
999,999 x 4 = 3,999,996
999,999 x 5 = 4,999,995
999,999 x 6 = 5,999,994
999,999 x 7 = 6,999,993
999,999 x 8 = 7,999,992
999,999 x 9 = 8,999,991

Notice how the digits in the ones and millions places always total nine.

Let me guess your age if you're over 9

Multiply your age by 10. From this product subtract 9 times any single digit you choose. Tell me the result, and I'll tell you your age.

?

EXAMPLE: Suppose your age is 17. Then 17x10=170. Suppose the digit you choose is 3. Then 3x9=27. And 170-27=143. The age is always the 2-digit number that appears in the hundreds & tens places plus the digit in the ones place. Here it is 14+3=17.

Let me quess

376 Pick any three digit number whose ones digit and hundreds digit are different.

673 Reverse its digits.

673-376=297
Find the difference between the two numbers.

792 Reverse its digits.

792+297=1089
Add the two together.
You ALWAYS get 1089!!!!

Puzzles to exercise your mind

These puzzles are meant to give your brain a workout.
So turn on your logic mode, and have fun at the same time.

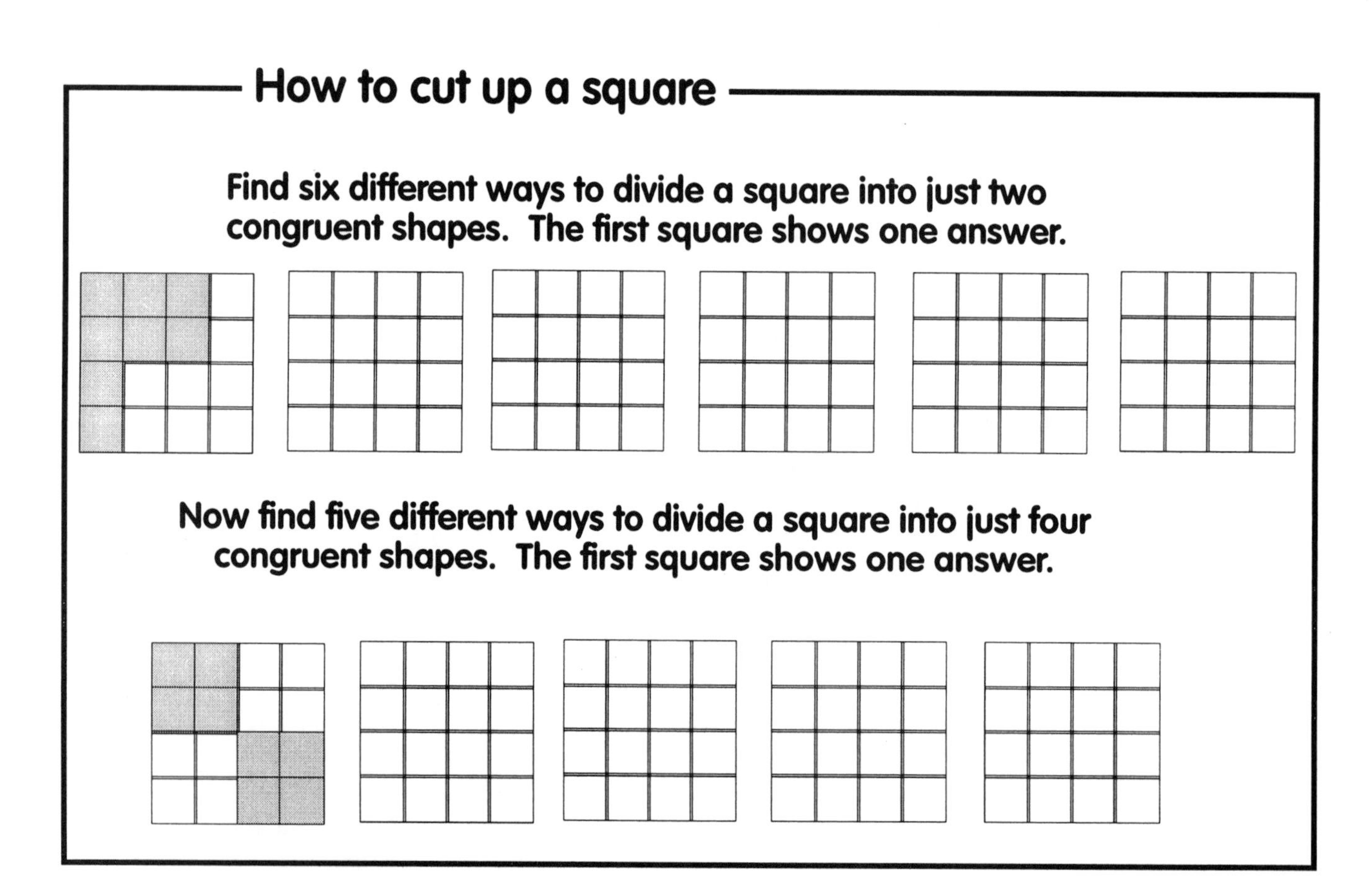

lines of coins

The coins are arranged so there are 3 rows of four coins in a row. Relocate only two coins and form 5 rows of four coins in each row.

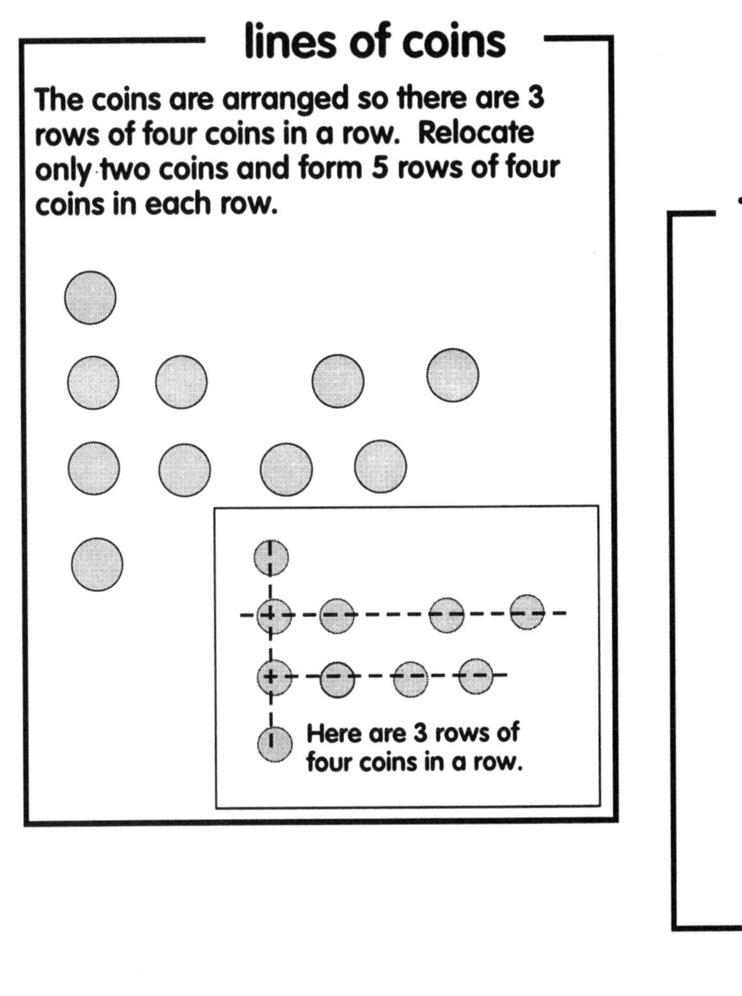

The operations puzzle

Using the digits from 1 through 9, place one digit in each box so all the equations are true.

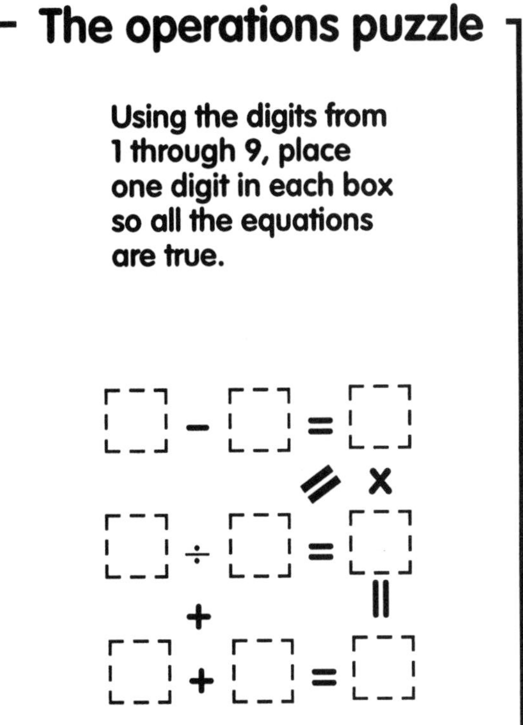

A study in eggs

This puzzle was created by the famous puzzlist Sam Loyd. How can a dozen eggs be arranged in this carton so that not more than two eggs are in each row, column, and diagonal?

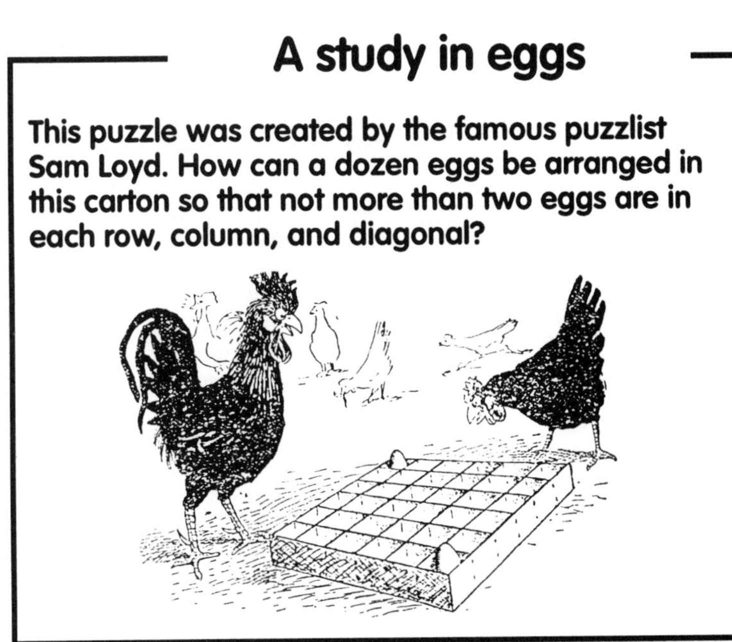

squares puzzle

Twenty-four toothpicks are arranged in squares. Remove 8 of them and leave a figure which has only two squares.

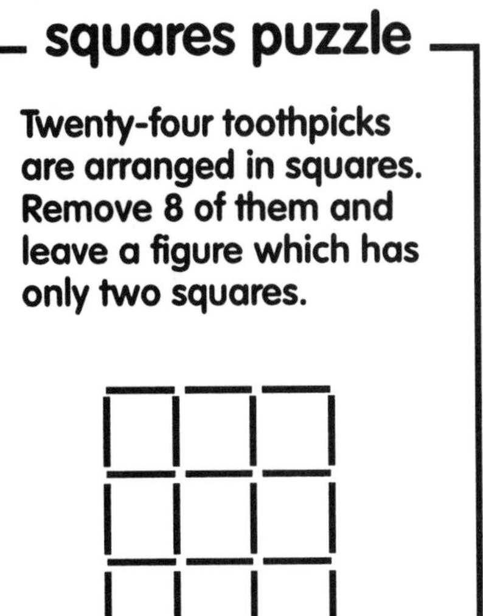

You can check your answers with those at the back of the book.

The game of grid

Grid is a game that can be played with pencil and paper or with stones.

It is an elimination game played by two people. Make a grid of dots, similar to this one.

Rules:

1) Use only horizontal or vertical line segments to connect dots.

2) Players take turns connecting dots. You may connect as many dots as you want during a turn, but you can only connect consecutive dots that are all vertical or all horizontal.

3) A player cannot draw a segment connecting dots over another sgement.

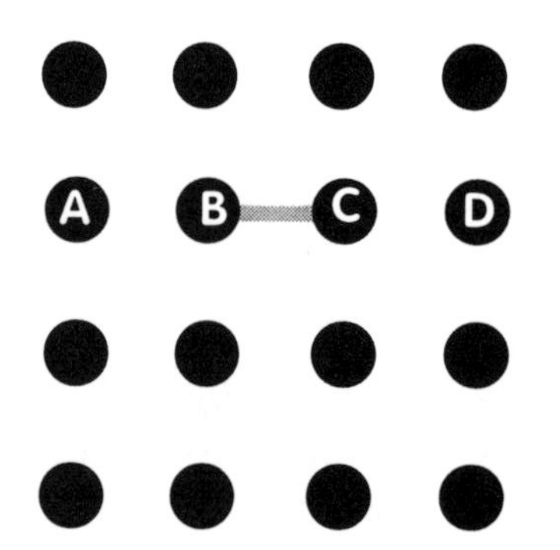

For example, a player drawing a horizontal segment from A cannot draw it to D because the segment would have to be drawn over segment BC.

4) The player who cannot connect any dots is the loser.

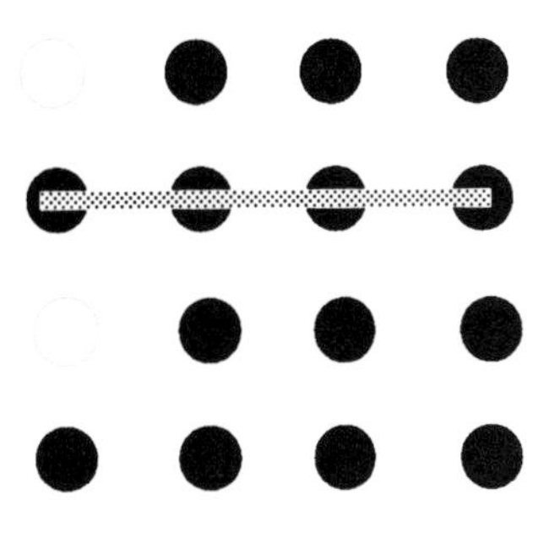

(1) First player's move.

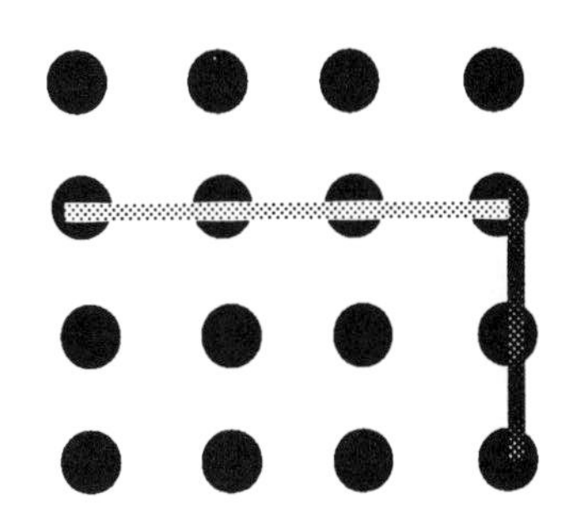

(2) Second player's move.

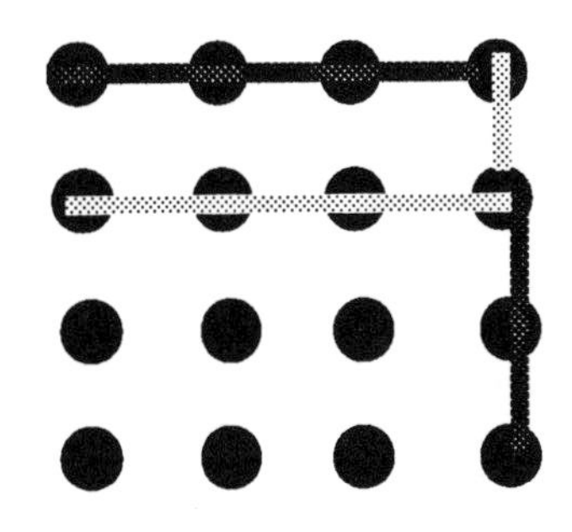

(3) Next two plays.

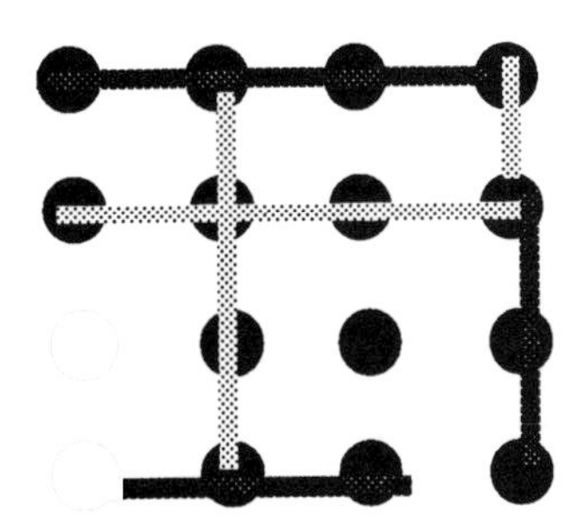

(4) Next two moves.

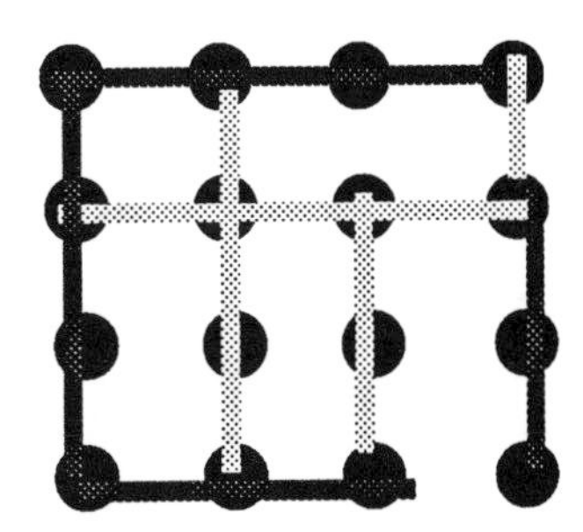

(5) Next two moves.

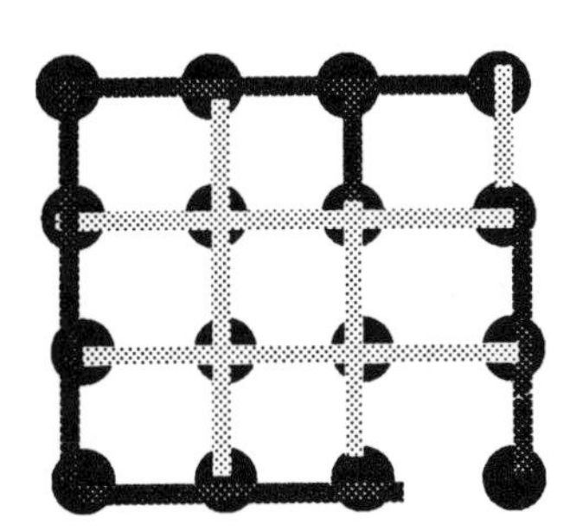

(6) Next two moves.

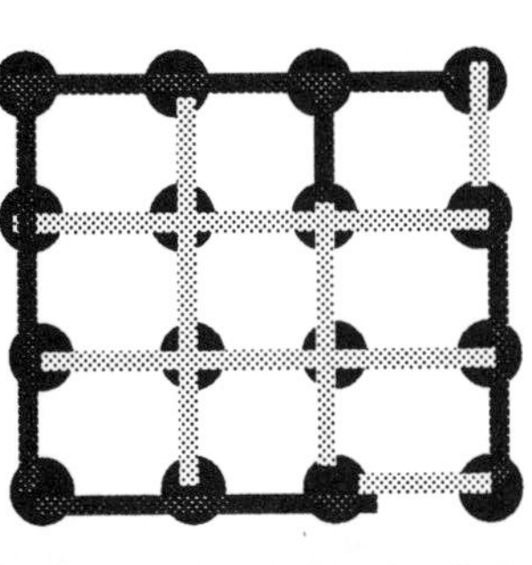

(7) The next play is for the light segment. It is the last possible move. Therefore, this player wins.

Use 3x3 or 4x4 grids for your first games. As you develop stategies, try them out on larger grids.

Try the game using stones or coins to make the grid. Then just remove coins from the line you would have drawn. Then the player to remove the last coin(s) is the winner.

What comes next?

Guessing what comes next can be both fun and challenging.

Study the groups of patterns carefully. Can you guess what comes next? The patterns get harder and harder. Take your time and don't come to hasty conclusions. The answers can be found at the end of the calendar. Try not to look until you've made a choice for each one.

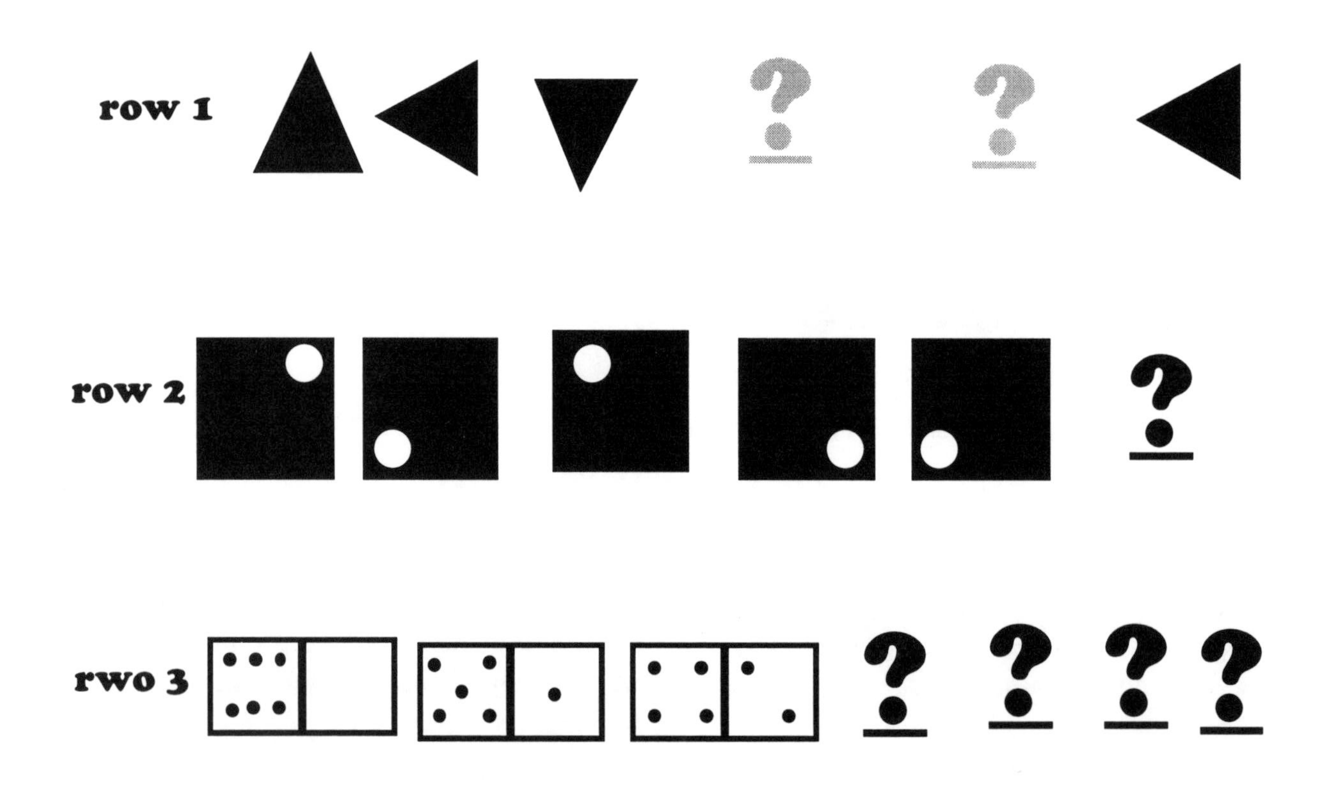

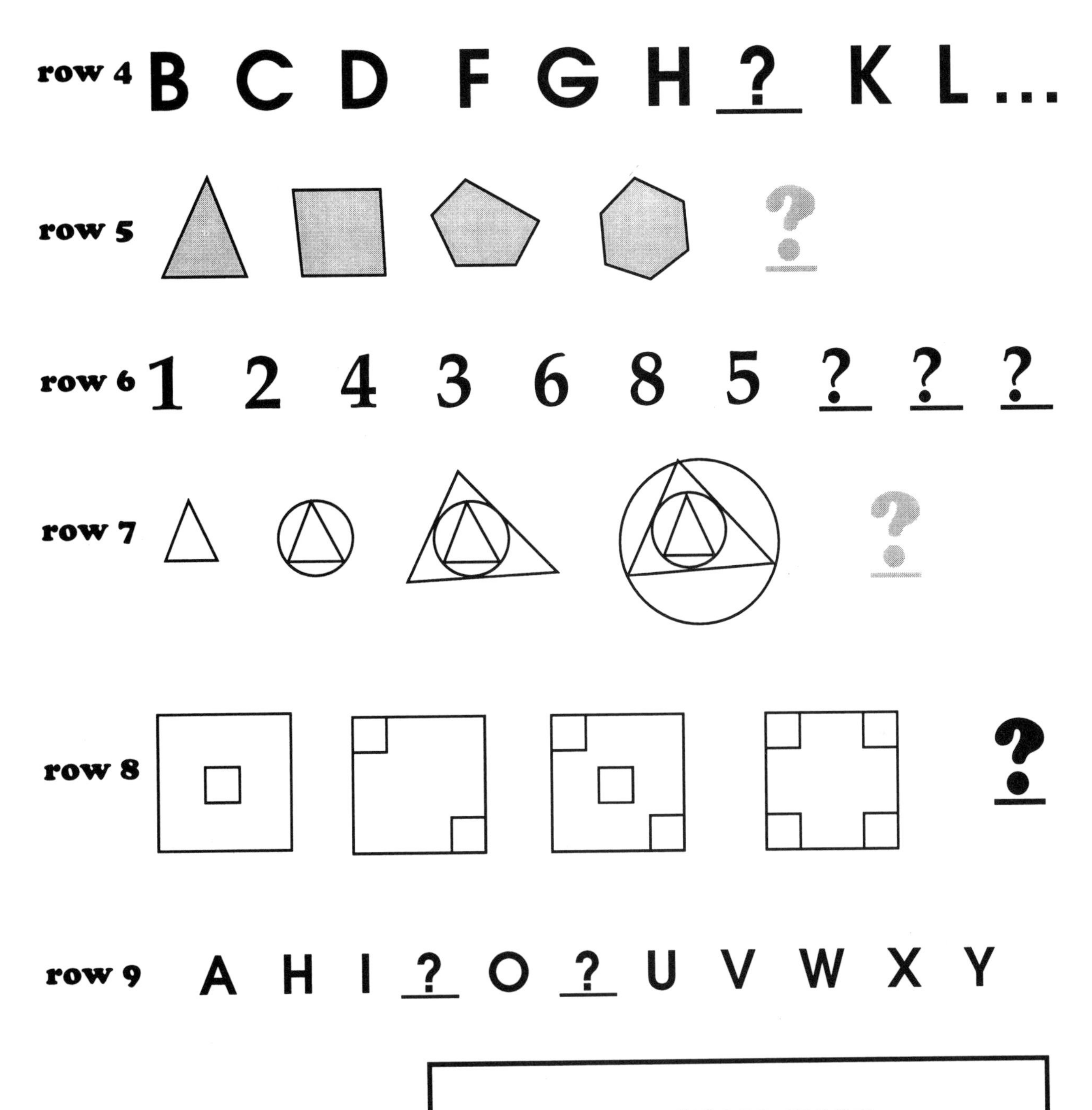

CHALLENGE

Make up your own logic patterns.

Test them out on a friend.

Brain busters

Here are some brain teasers that may bust your brains.

You can check your answers with those at the back of the book.

Hypercard

Study the drawing of a card that has been cut and folded. Using a rectangular shaped card, can you duplicate it?

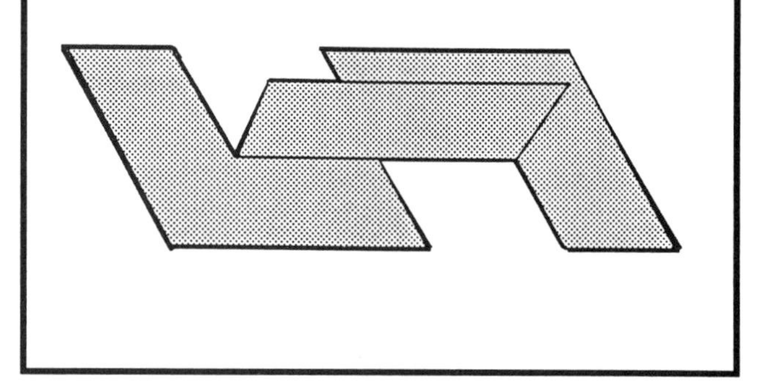

The penny puzzle

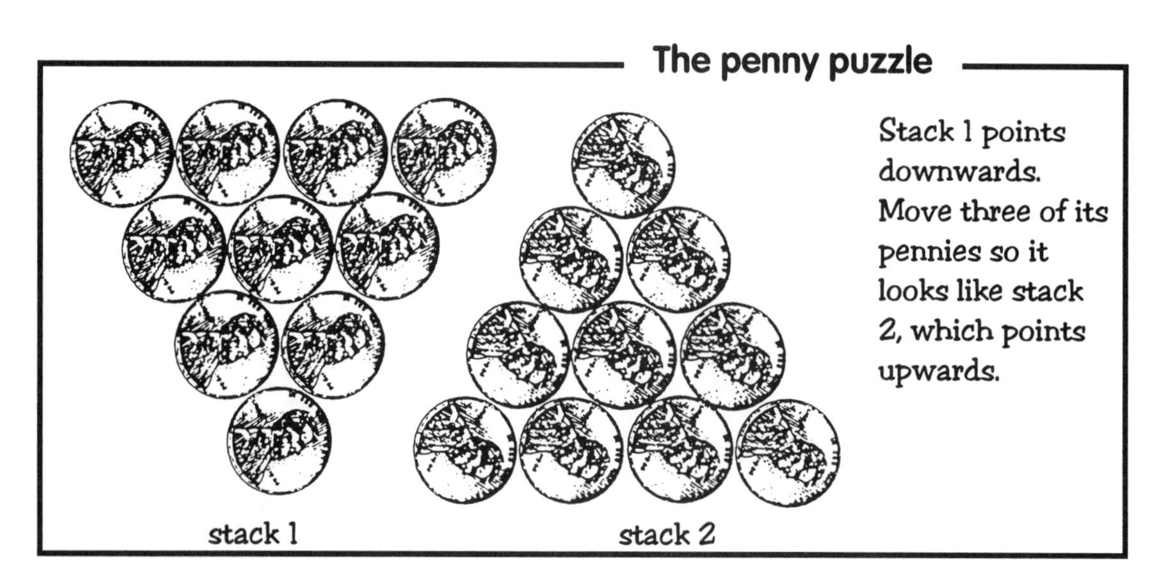

Stack 1 points downwards. Move three of its pennies so it looks like stack 2, which points upwards.

Sam Loyd's jockey puzzle

Make a photocopy of the drawing below. Cut the picture into three rectangles along the dotted lines. Rearrange the rectangles without folding them to show two jockeys riding two galloping donkeys.

HINT: Review the story of *The Persian Horses.*

Sam Loyd's sheep puzzle

The farmer wants to fence off each sheep in its own pen. Using only three straight fences, how can the farmer do this?

The boat puzzle

A farmer needs to take his goat, wolf and cabbage across the river. His boat can only take him and either his goat, wolf, or cabbage. If he takes the wolf with him, the goat will eat the cabbage. If he takes the cabbage, the wolf will eat the goat. Only when the farmer is present is the cabbage safe from the goat and the goat safe from the wolf. **How does he get everything across the river?**

Hidden figures puzzle

This figure is made from a square and four equilateral triangles.

A triangle is equilateral if all of its sides are the same length.

The equilateral triangles are placed at the 4 corners (called vertices) of the square.

Many geometric shapes in various sizes are hidden in this figure. Can you find at least–

4 squares
16 equilateral triangles
12 isoscles triangles
an octagon
8 right triangles
4 trapezoids
8 pentagons
4 hexagons
4 parallelograms
16 quadrilaterals

isosceles right triangle
isosceles triangle
hexagon
octagon
pentagon
trapezoid
qudrilateral
parallelogram
right triangle

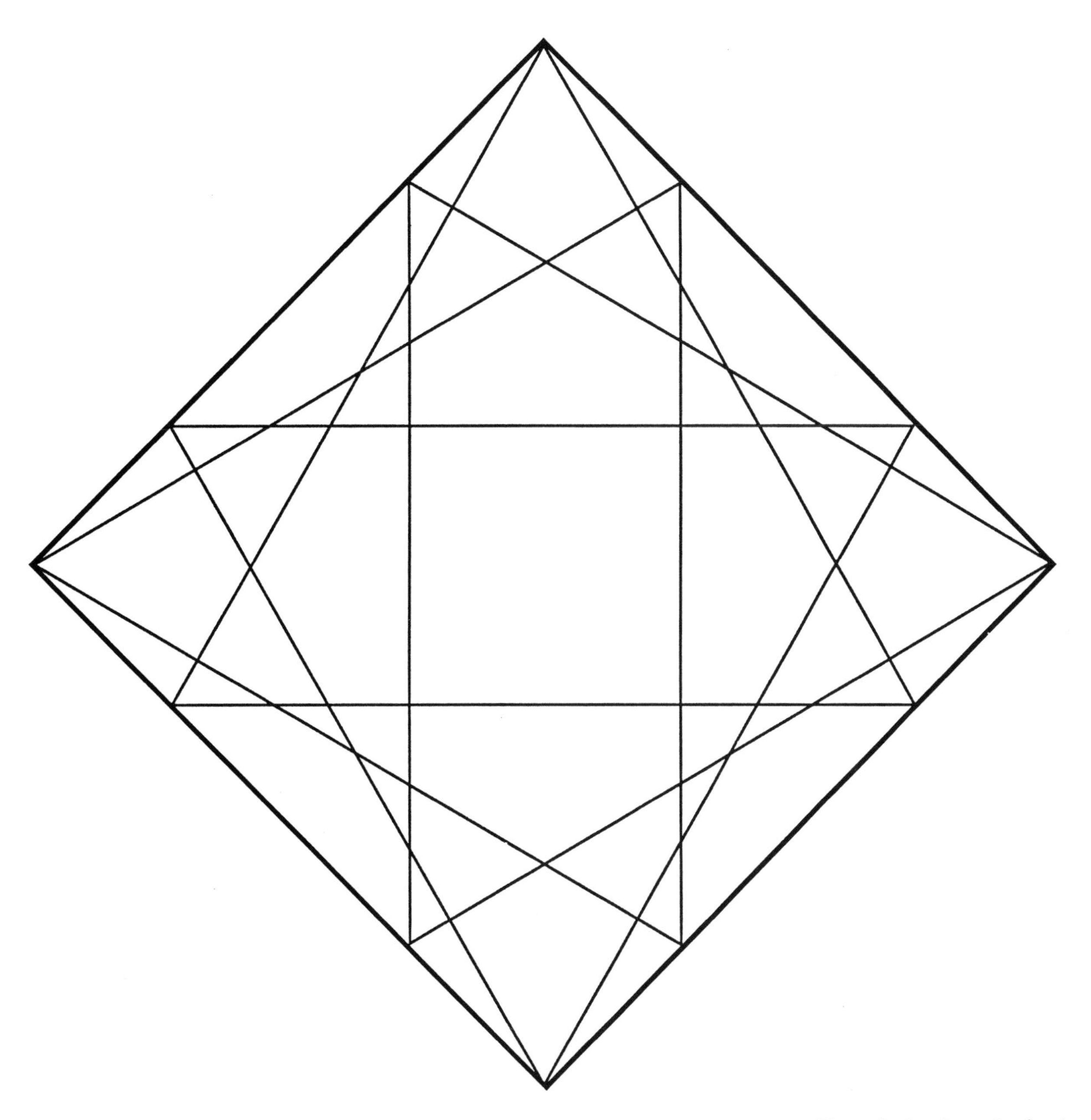

The solution is at the back of the book.

Discovering number patterns

When you first begin to study mathematics, the main characters are numbers. Some numbers have properties or do things that almost seem magical. Here are some wild number patterns for you to discover. You will need a pencil, paper and probably a hand calculator.

Study the numbers and notice how they are changing. For each a pattern or patterns will appear. Fill in the missing numbers as you figure out what the patterns are. Don't actually do the computation. After you have made your guess, check out what you guessed by computing the calculations by hand or calculator.

1,089 x 1	= 1,089
1, 089 x 2	= 2,178
1, 089 x 3	= _, 267
1, 089 x _	= 4,356
1, 089 x _	= _ , _ _ _
1, 089 x _	= _ , _ _ _
1, 089 x _	= _ , _ _ _
1, 089 x _	= _ , _ _ _
1, 089 x _	= _ , _ _ _

Answers are given in the solution section at the end of the book.

0 x 9	+8	= 8
9 x 9	+7	= 88
98 x 9	+6	= _ _ _
987 x 9	+ 5	= _ , _ _ _
_ , _ _ _ x 9	+ _	= _ _ , _ _ _
_ _ , _ _ _ x 9	+ _	= _ _ _ , _ _ _
_ _ _ , _ _ _ x 9	+ _	= _ , _ _ _ , _ _ _
_ , _ _ _ , _ _ _ x 9	+ _	= _ _ , _ _ _ , _ _ _

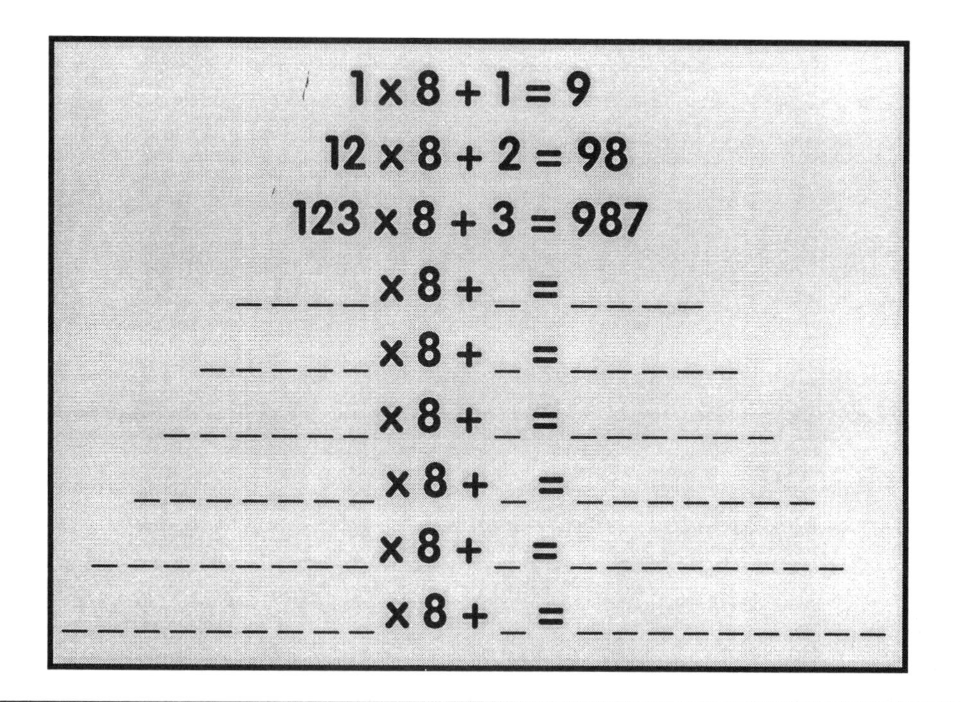

495 is another really tricky number

1) Pick any three different digits (0,1,2,3,4,5,6,7,8, or 9).

2) Using the three digits you picked, put them together to make the largest three digit number. For example, if you picked 3, 1, and 8 then the largest three digit number would be 831. Now make the smallest three digit number from the digits you picked, 138.

3) Subtract these two numbers. 831 - 138 = 693

4) Now using the three digits from this difference, repeat steps 2 and 3.

963 - 369 = 594

Continually repeat steps. What happens?

Eventually you will get 495, and when you do steps 2 and 3 to 495 you get 495 again. Pretty tricky!

Try the same process, but this time picking four digits. Do you get a tricky number, and if so what number?

Mental push-ups

Here are some puzzle problems to exercise your mind.

The hat problem

In a box there are 1 black hat and 2 tan hats. Tom and Jerry are blindfolded, and a hat from the box is placed on each boy's head. The blindfolds are removed, and they are asked –
"Who knows what color hat they are wearing?"
Tom saw that Jerry had on a tan hat, and exclaimed in frustration, "How could I know!" Then Jerry with a smug voice said, "Well, I know what color I'm wearing."
How could Jerry know what color he was wearing? What color did he see on Tom's' head?

Arranging the line-up

By touching only checkers A and B arrange the checkers so the columns are all the same color checkers.

The small change problem

This piggy bank has $1.05 in coin change. Figure out which coins are in the bank, if you know that:

a) The bank does not have the right coins to change a nickel.

b) The bank does not have the right coins to change a dime.

c) The bank does not have the right coins to change a quarter.

d) The bank does not have the right coins to change a half dollar.

e) The bank does not have the right coins to change a dollar.

f) None of the coins is a dollar coin.

How did the digits get there?

0 2 ?

1 3 ?

- Ask a friend to pick any three digits from 0 to 9.
- Ask your friend to add 4 to the first choice, and then multiply this sum by 10.
- Now have your friend add the second choice to this product, and multiply this by 10.
- To this add the third digit picked.
- Finally, subtract 400.

The final number is made up of the three digits your friend picked– all in the order picked!

How does this work?

Solutions are given at the back of the book.

The game of Awithlaknannai

Awithlaknannai is a fast moving capture game played by the Zuni Indians of New Mexico.

You can make a formal playing board, draw one on paper, or on the earth in your yard. All you need is a friend and 12 playing pieces for each player. You can use anything you like for pieces – peebles, paper clips, jelly beans, pennies (with head & tails for each player), checkers.

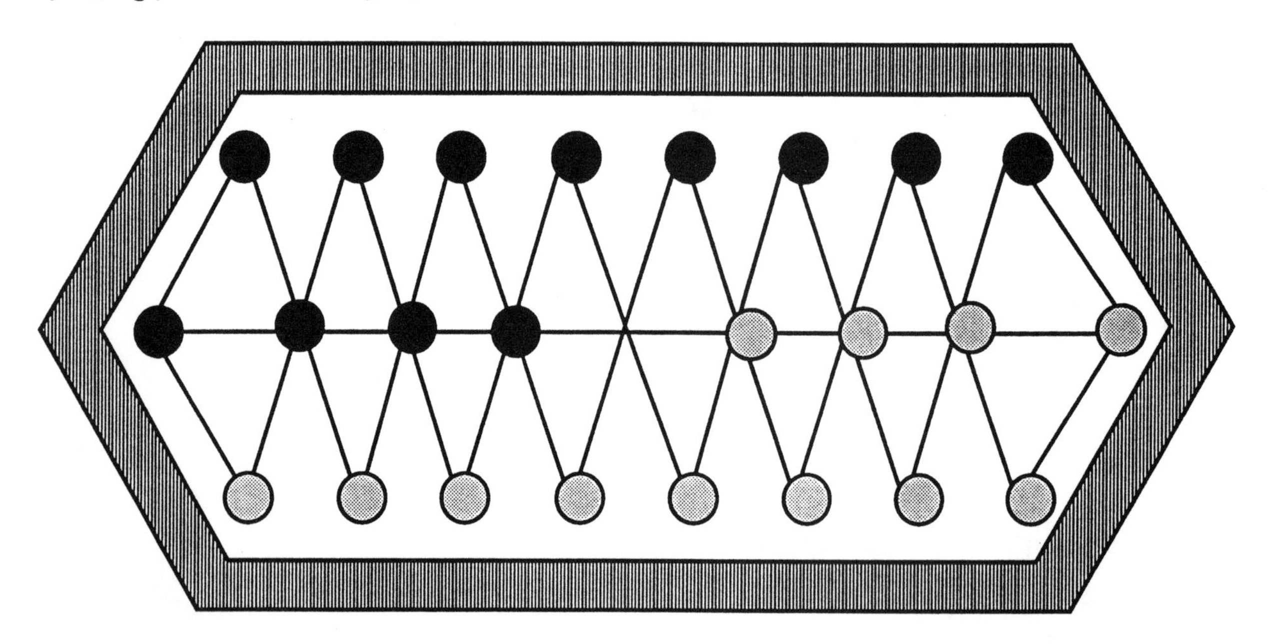

How to play.

1) Set up your pieces as shown in the start up position in the illustration.

2) Players alternate turns moving along the lines.

3) A player can move one space to an empty adjacent space in any direction.

4) Pieces are *captured* by jumping an opponent's adjacent piece and landing on an empty space next to the piece being captured. You can capture as many pieces as possible on your turn.
For example, the black piece with the

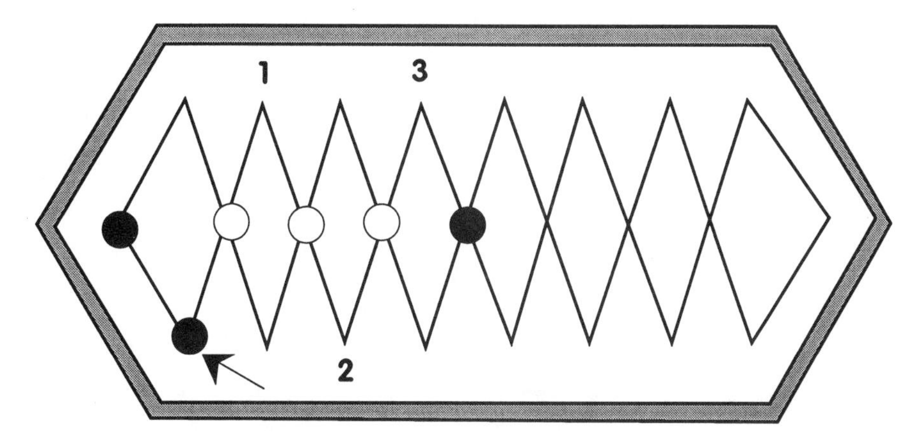

arrow can jump 3 times on its turn and capture 3 pieces on this turn by going to spaces 1, 2, and 3.

5) The winner is the player who captures all of his opponent's pieces.

Mind bending puzzles

Here are puzzles to bend your brain. Put on your logic hat, and have fun at the same time. Enjoy!

This puzzle was created by Sam Loyd (1841-1911), one of America's famous puzzle maker. He started his puzzle career when he was a teenager, when he was problem editor for a chess magazine.

Find the five pointed star hidden in this figure.

the penny puzzle

Draw three straight lines in the square so that each penny is separated by a lines from the other pennies.

getting the squares puzzle

Remove 5 toothpicks and end up with three squares the same size.

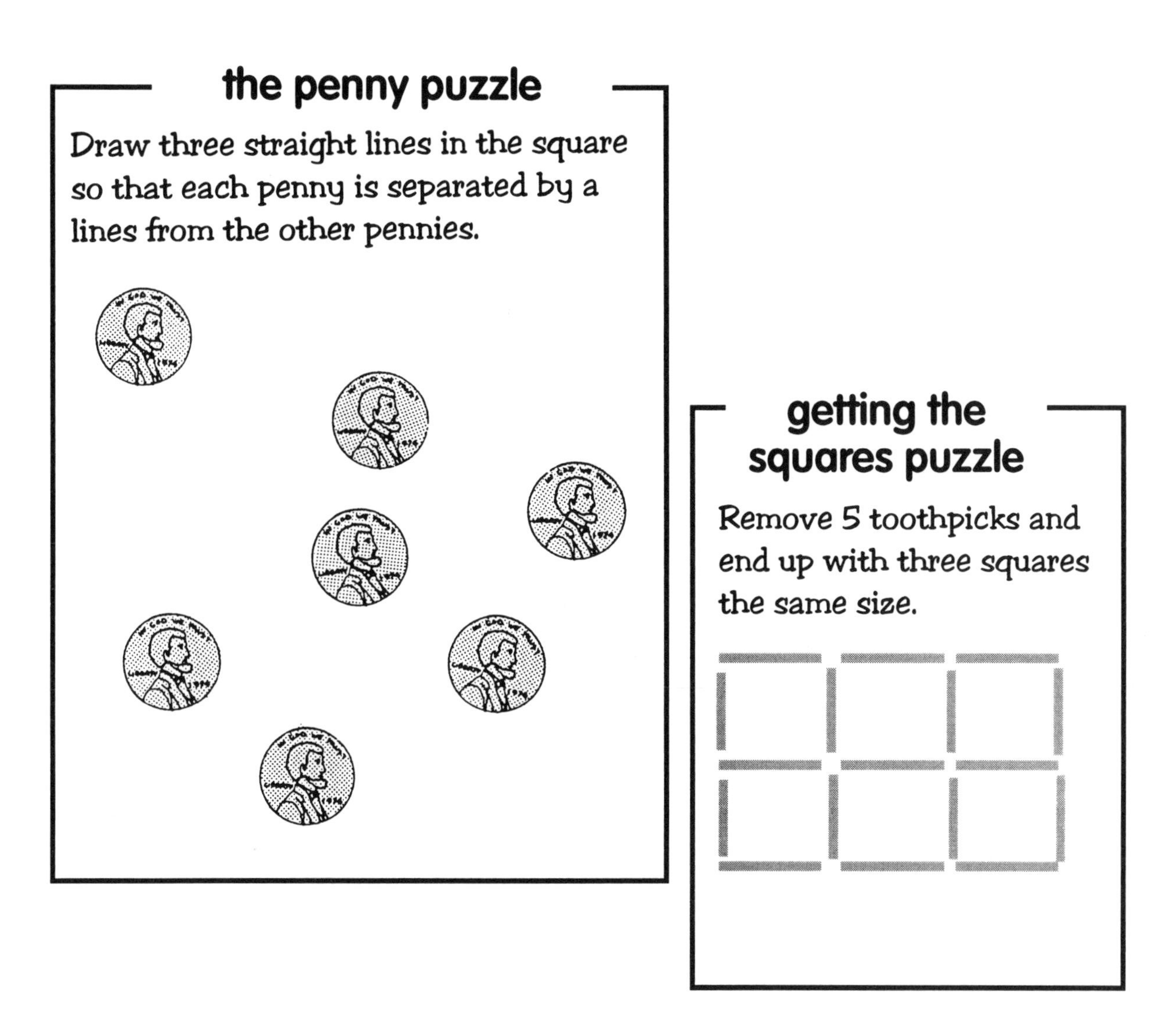

How to cut the cake

Using only three straight cuts, find the most number of pieces you can cut this round cake.

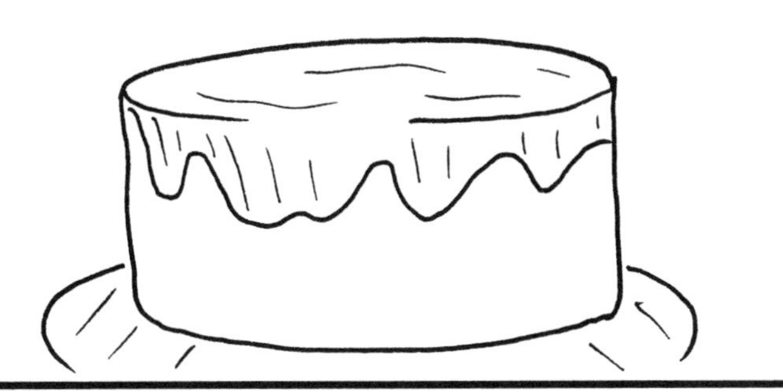

You can check your answers with those at the back of the book.

Where's the missing object?

Ever lose something, and not been able to figure out where it is?

Well, here are some lost object puzzles. See if you can figure out where each thing went.

Where is the missing dollar? Three friends ordered 3 giant ice cream sundaes with 3 scoops each. They each paid $5. Later the clerk remembered the 3 for $10 special, and took $5 from the register. He gave each friend $1 and kept $2. So each friend paid $4 for a total of $12, and the clerk kept $2. That totals $14. Where is the other $1 from the $15?

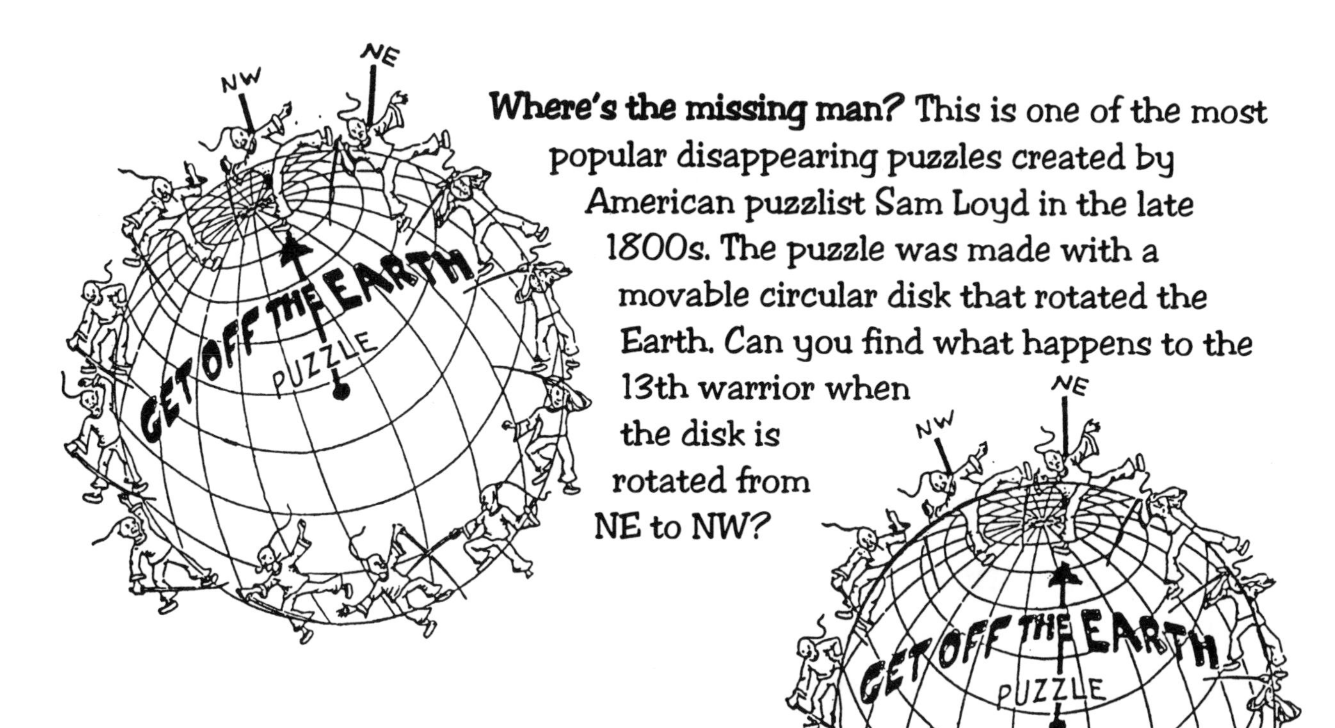

Where's the missing man? This is one of the most popular disappearing puzzles created by American puzzlist Sam Loyd in the late 1800s. The puzzle was made with a movable circular disk that rotated the Earth. Can you find what happens to the 13th warrior when the disk is rotated from NE to NW?

Where is the missing desk? The school library had *six* unoccupied reading desks available today. Tom and *six* other students came in at the same time to read at these desks. The librarian thought for a moment and told Tom to wait, while she seated the other students. She placed the 2nd student at the 1st desk, the 3rd student at the 2nd desk, the 4th student at the 3rd desk, the 5th student at the 4th desk, and the 6th student at the 5th desk. She went back and led Tom to the 6th desk. How did she manage to place all 7 students at 6 desks?

Putting on your logic hat

Logic, a way of thinking, is a very important part of mathematics.

Here are some logic questions that are fun, and also make you think twice.

Try them out on yourself, then ask them of a friend. Careful, some of them are pretty tricky.

1) How many birthdays does the average person have?

2) Do they have a fourth of July in England?

3) Why can't a man living in San Francisco be buried in New York city?

4) How far can a large dog run into the woods?

5) Two men play five games. Each man wins the same number of games. How do you explain this?

6) Suppose I have only one match. I enter a room where there was an oil lamp, an oil heater, a candle, and kindling wood. Which would I light first?

7) The doctor gave me three pills and told me to take one every 30 minutes. How long would the pills last?

8) The archaeologist found a gold coin marked 46 B.C. How did she know the coin was fake?

9) In baseball, how many outs in an inning?

10) A gave Mary a vaccination. Mary is the nurse's sister, but the nurse is not the Mary's sister. Why?

11) At the same time, Sarah stands behind Tim and Tim stands behind Sarah. How can they do this?

12) Which weighs more? A pound of tennis balls, or a pound of rocks?

13) Is it legal for a man in Oregon to marry his widow's sister?

14) The first 15 days of this month it rained all but 8 days. How many days did it not rain?

15) Paul's little brother and sister are playing in a sand box. His brother made 3 sand piles, while his sister made 5 piles. If they put their sand piles together, how many sand piles will there be?

16) Take two apples from five apples, what do you have?

17) How many cubic inches of earth are in a hole the shape of a box that is 2" wide, 3" long and 4" deep?

18) In my hand I have only two U.S. coins which total 26 cents. One is not a penny. What are they?

19) Some months have 30 days, some have 31 days. How many months have 29 days during a leap year?

The answers are written in the solutions section at the end of the book.

A twist to the Möbius strip

Mathematics has some very unusual objects, which hold some very interesting surprises.

Objects like these two are studied in an area of mathematics called topology. Topology studies what properties of objects remain unchanged when certain things are done to them. Make these paper models and some big surprises will be in store for you.

Here's how to make **model A.**

1) Take a sheet of paper.

2) Cut out a rectangle from each corner, so you have left this shape.

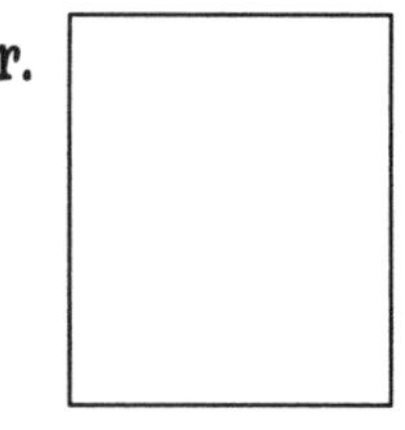

3) Take A and B fold them back, so they make a ring, like this and tape strips A and B together.

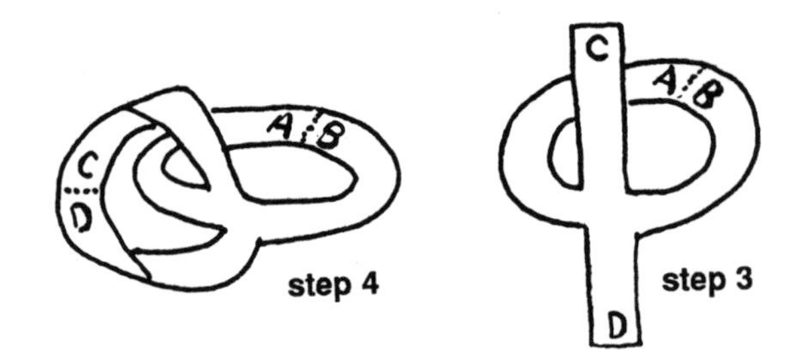

4) Now take strip C give it a half twist and tape it to strip D.

5) Cut along an imaginary dotted line, as in the large model, and see how the object changes.

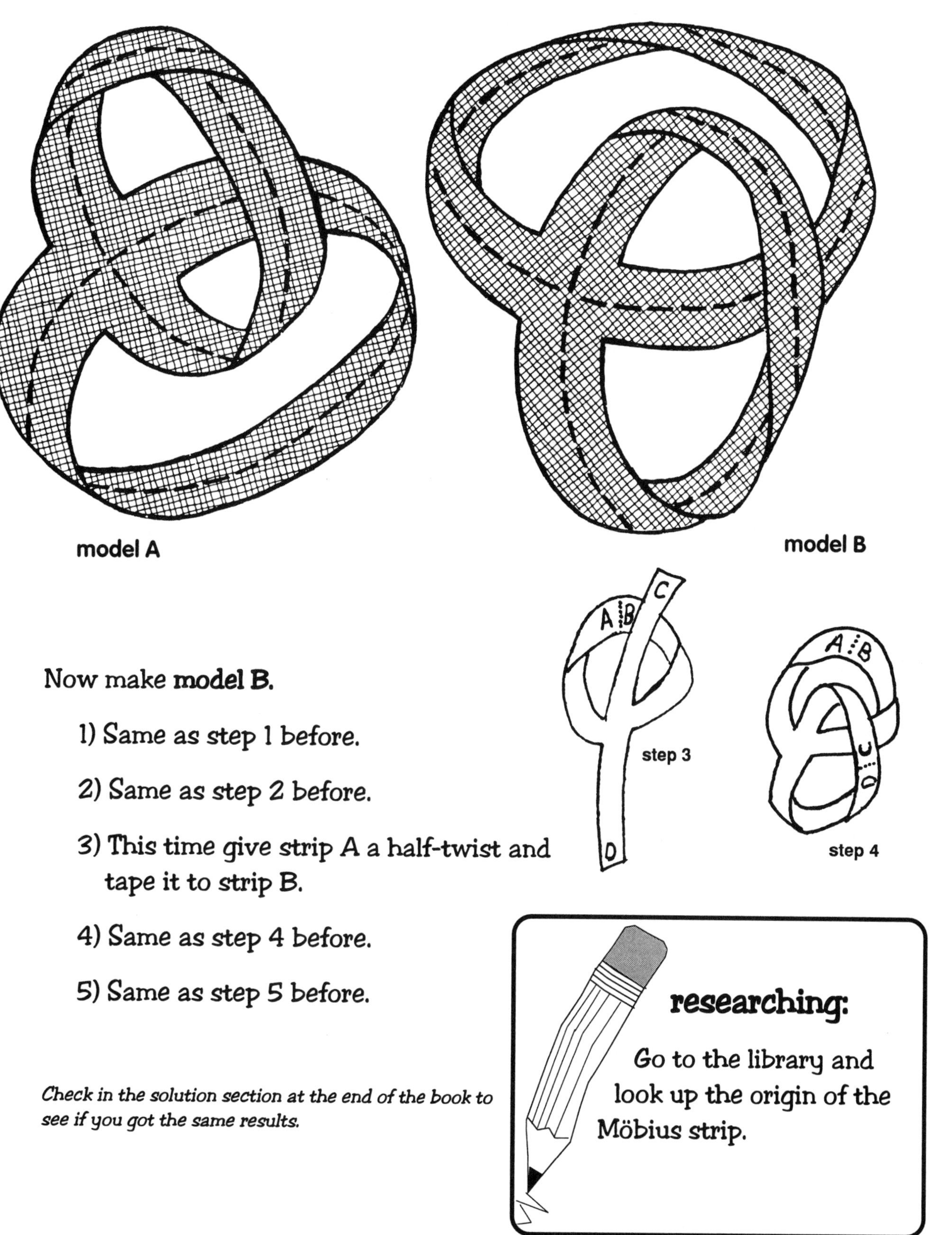

Now make **model B.**

1) Same as step 1 before.

2) Same as step 2 before.

3) This time give strip A a half-twist and tape it to strip B.

4) Same as step 4 before.

5) Same as step 5 before.

Check in the solution section at the end of the book to see if you got the same results.

researching:

Go to the library and look up the origin of the Möbius strip.

The game of Chase

CHASE is an easy game to construct, and a lot of fun to play. You'll need a large piece of cardboard, a sheet of stiff or card like paper, a ruler, and a dark marking pencil or pen. With the sheet of cardboard make a rectangle that is 8"x16". With your ruler and pen, divide the rectangle into 144 one inch squares, as shown in the illustration. This will be your playing board.

Next, make the playing pieces with the piece of stiff paper. Each of the two players will have 24 playing pieces–8 circles, 8 triangles, and 8 squares–that will fit on the squares of the playing board. Shade one of the sets of playing pieces, so the two players can distinguish their pieces. Now you're ready to play.

About the playing pieces

The *circle* moves only one space at a time in any direction.

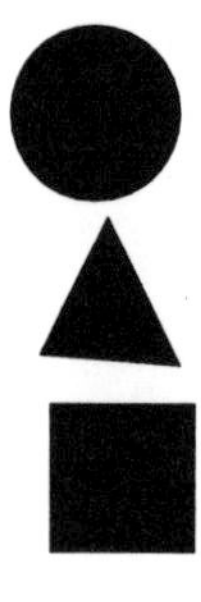

The *triangle* must move 3 consecutive spaces at a time in any direction.

The *square* must move 4 consecutive spaces at a time in any direction.

Moving

The pieces can move forwards, backwards, or sideways. Pieces can jump over other pieces when moving, but must count the jumped over piece as a space moved.

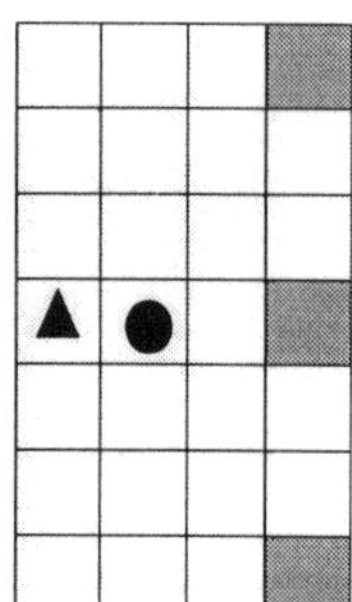

The triangle can move 3 spaces to any of the dark shaded squares.

How to capture pieces

If on your turn you can move your piece so that it lands on your opponent's piece, then you take your opponent's piece and your piece now occupies that place on the board.

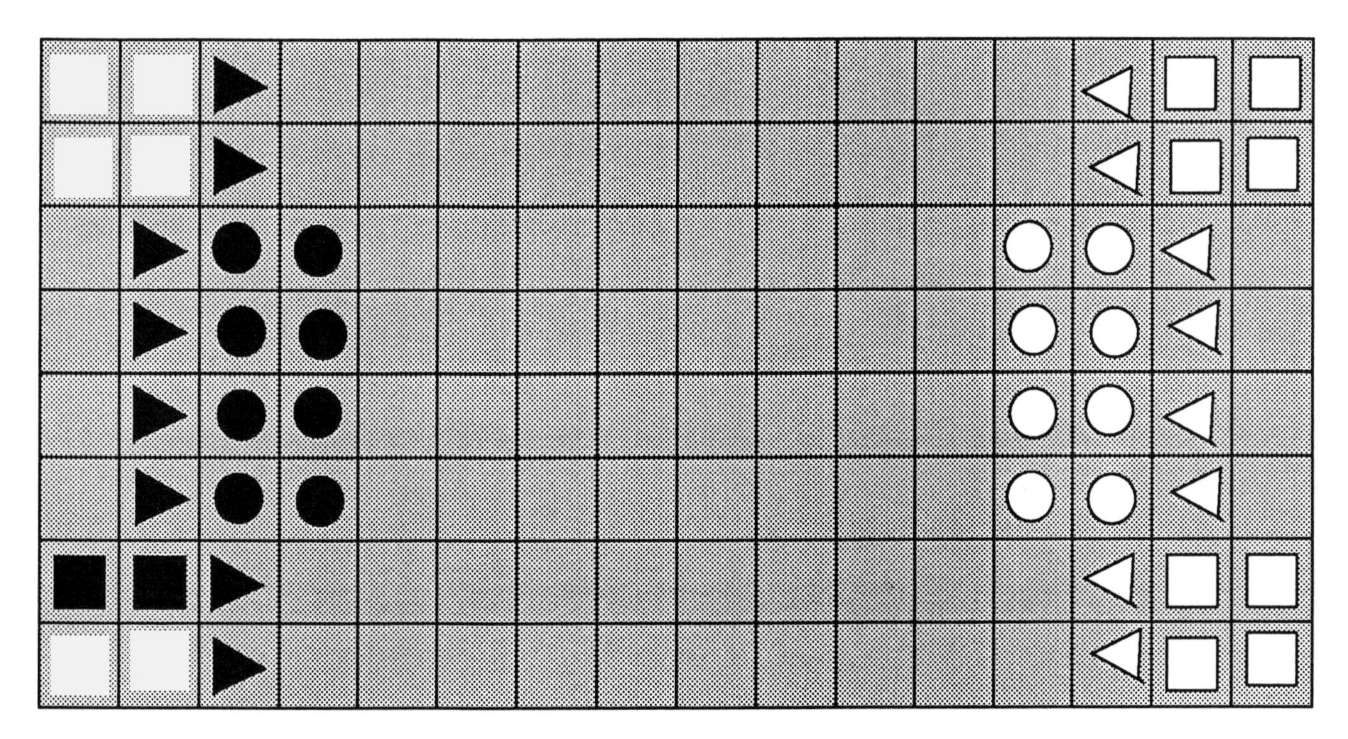

The starting positions of the pieces.

The object of the game

Before beginning to play, the players decide how long they want the game to last by deciding how many captured pieces make a win. You may want to start off making the winner the first to captures 10 of the other player's pieces.

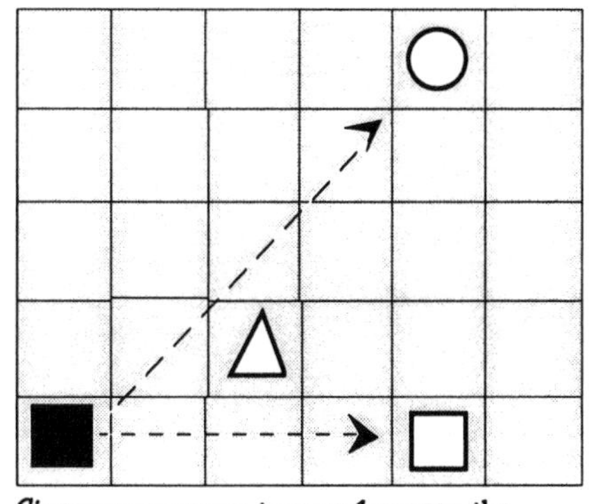

Since a square must move 4 spaces, the black square can move diagonally 4 spaces and capture the white circle or move 4 spaces straight ahead and capture the white square.

This is the basic way to play CHASE. There are many variations of the game. It's simple to learn, but challenging to play. Have fun!

CHASE was developed by Theoni Pappas from an old game called Rithmomachia.

solutions & answers section

answers & solutions

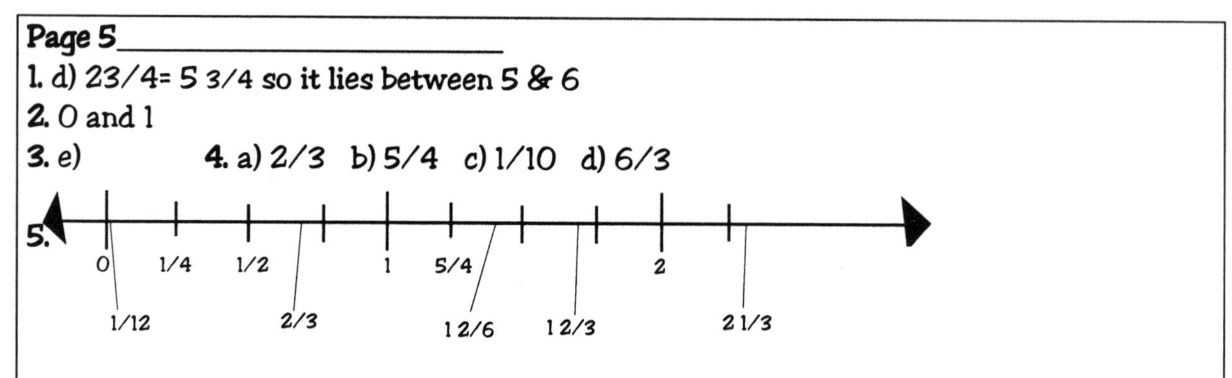

Page 5

1. d) 23/4= 5 3/4 so it lies between 5 & 6
2. 0 and 1
3. e)

4. a) 2/3 b) 5/4 c) 1/10 d) 6/3

5.

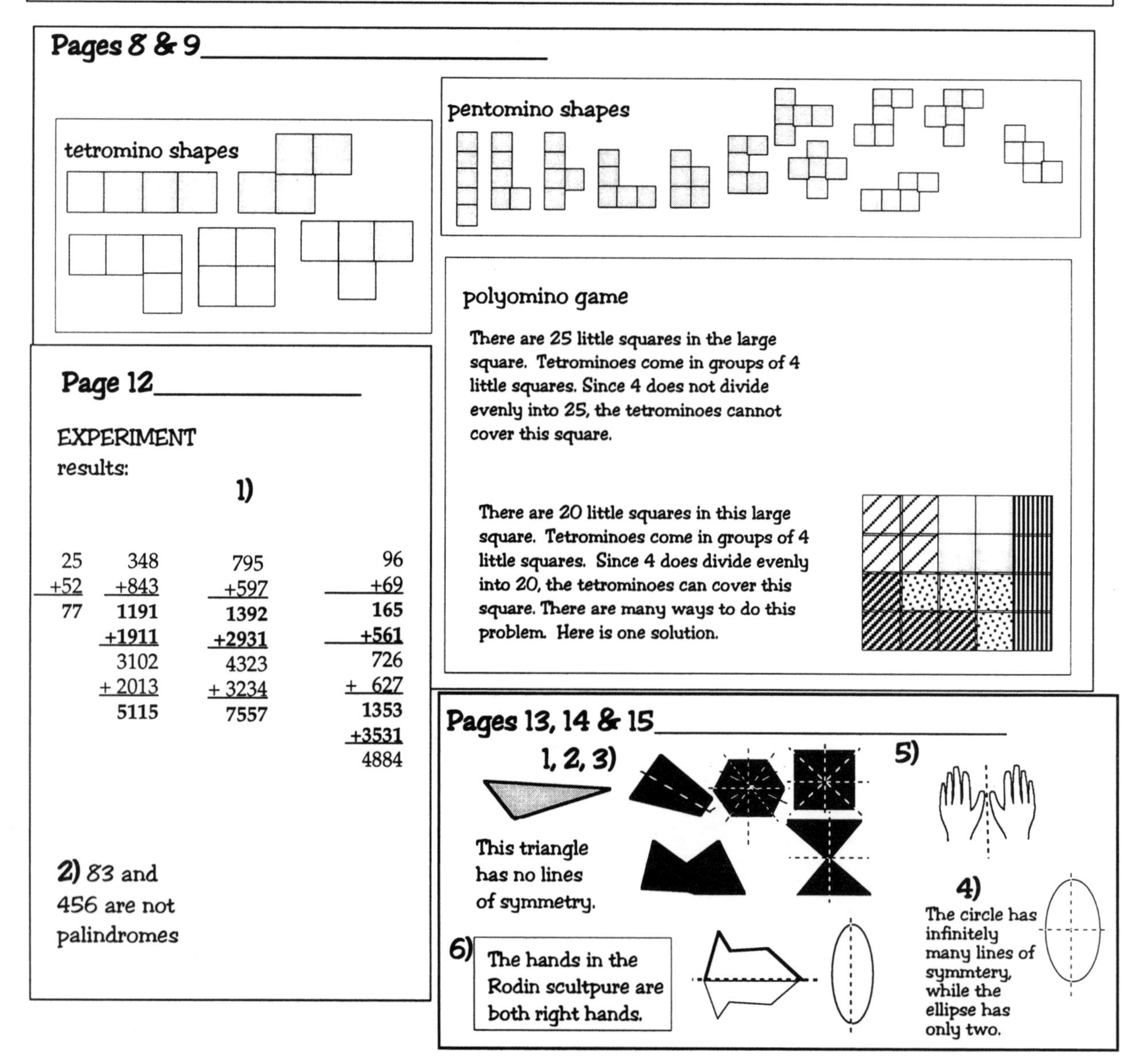

Pages 8 & 9

polyomino game

There are 25 little squares in the large square. Tetrominoes come in groups of 4 little squares. Since 4 does not divide evenly into 25, the tetrominoes cannot cover this square.

There are 20 little squares in this large square. Tetrominoes come in groups of 4 little squares. Since 4 does divide evenly into 20, the tetrominoes can cover this square. There are many ways to do this problem. Here is one solution.

Page 12

EXPERIMENT results:

1)

```
 25     348     795      96
+52    +843    +597     +69
 77    1191    1392     165
      +1911   +2931    +561
       3102    4323     726
      +2013   +3234    +627
       5115    7557    1353
                      +3531
                       4884
```

2) 83 and 456 are not palindromes

Pages 13, 14 & 15

1, 2, 3)

This triangle has no lines of symmetry.

4) The circle has infinitely many lines of symmtery, while the ellipse has only two.

5)

6) The hands in the Rodin scultpure are both right hands.

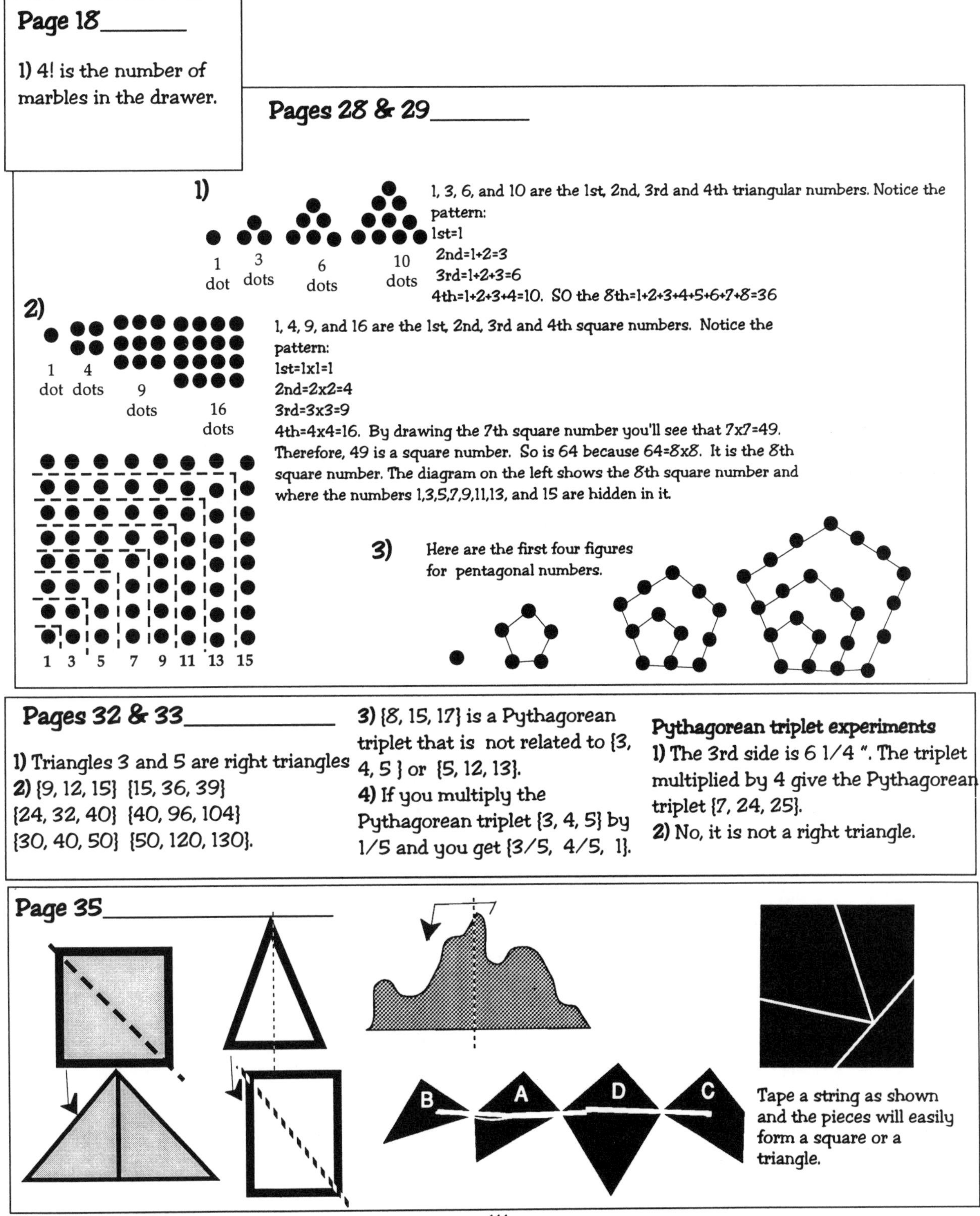

Page 18

1) 4! is the number of marbles in the drawer.

Pages 28 & 29

1) 1, 3, 6, and 10 are the 1st, 2nd, 3rd and 4th triangular numbers. Notice the pattern:
1st=1
2nd=1+2=3
3rd=1+2+3=6
4th=1+2+3+4=10. SO the 8th=1+2+3+4+5+6+7+8=36

2) 1, 4, 9, and 16 are the 1st, 2nd, 3rd and 4th square numbers. Notice the pattern:
1st=1x1=1
2nd=2x2=4
3rd=3x3=9
4th=4x4=16. By drawing the 7th square number you'll see that 7x7=49. Therefore, 49 is a square number. So is 64 because 64=8x8. It is the 8th square number. The diagram on the left shows the 8th square number and where the numbers 1,3,5,7,9,11,13, and 15 are hidden in it.

3) Here are the first four figures for pentagonal numbers.

Pages 32 & 33

1) Triangles 3 and 5 are right triangles
2) {9, 12, 15} {15, 36, 39}
{24, 32, 40} {40, 96, 104}
{30, 40, 50} {50, 120, 130}.
3) {8, 15, 17} is a Pythagorean triplet that is not related to {3, 4, 5 } or {5, 12, 13}.
4) If you multiply the Pythagorean triplet {3, 4, 5} by 1/5 and you get {3/5, 4/5, 1}.

Pythagorean triplet experiments
1) The 3rd side is 6 1/4 ". The triplet multiplied by 4 give the Pythagorean triplet {7, 24, 25}.
2) No, it is not a right triangle.

Page 35

Tape a string as shown and the pieces will easily form a square or a triangle.

answers & solutions

Page 40

1) 23, and 31 are not multiples of 3.

2) 30, 50 and 100 are multiples of 10.

3) 4, 8, 12, 16, 20, 24, 28 are the multiples of 4 which are less than 32.

4) 20, 25 and 30 are multiples of 5 between 15 and 35.

5) 12 itself is the smallest multiple of 1, because 1x12=12.

6) 195 is the largest multiple of 5 which is less than 200.

Page 42

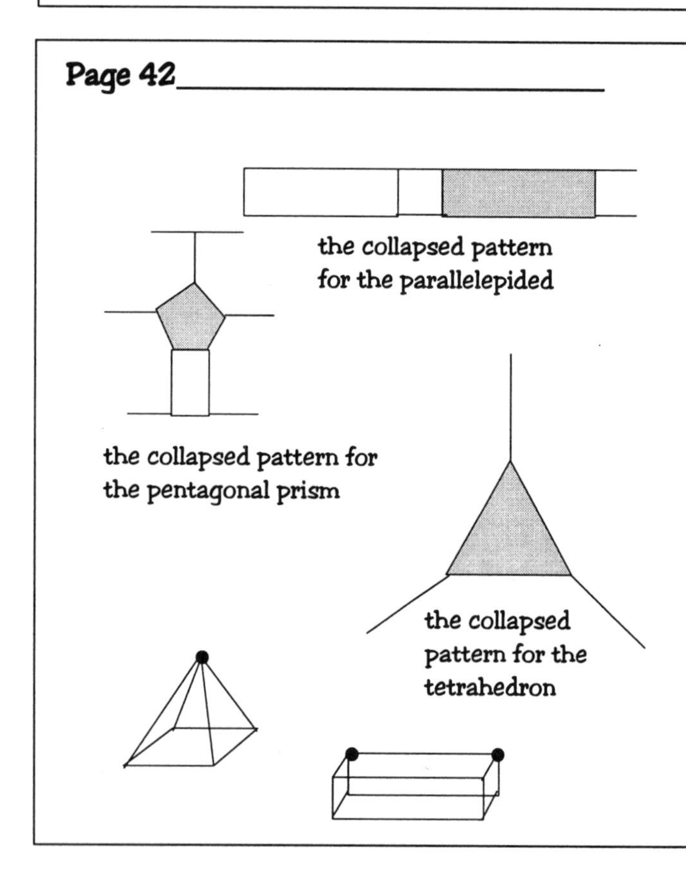

Page 45

0.142857… = 1/7

0.3333333… = 1/3

25% = 1/4

0.125 = 1/8

0.1 =1/10

10/40 = 1/4

725/800 = 29/32

1% = 1/100

0.777 = 7/9

0.07 =7/100

Page 36 & 37

THE BARBER PARADOX
If the barber shaves himself, then the barber is shaving someone who shaves himself. But if the barber does not shave himself, then the barber is supposed to shave himself. Aren't we going around in circles?

THE SENTENCE PARADOX
How do you know not to read it if you don't read it?

THE MISSING BAR PARADOX
A bar disappears when it becomes a side in the top half of the rectangle.

THE IMPOSSIBLE FIGURE PARADOX
When looking at this, the mind is confused by an impossible figure. It cannot really exist. If your hand covers up the lower part, the figure becomes realistic.

answers & solutions

Pages 62 & 63

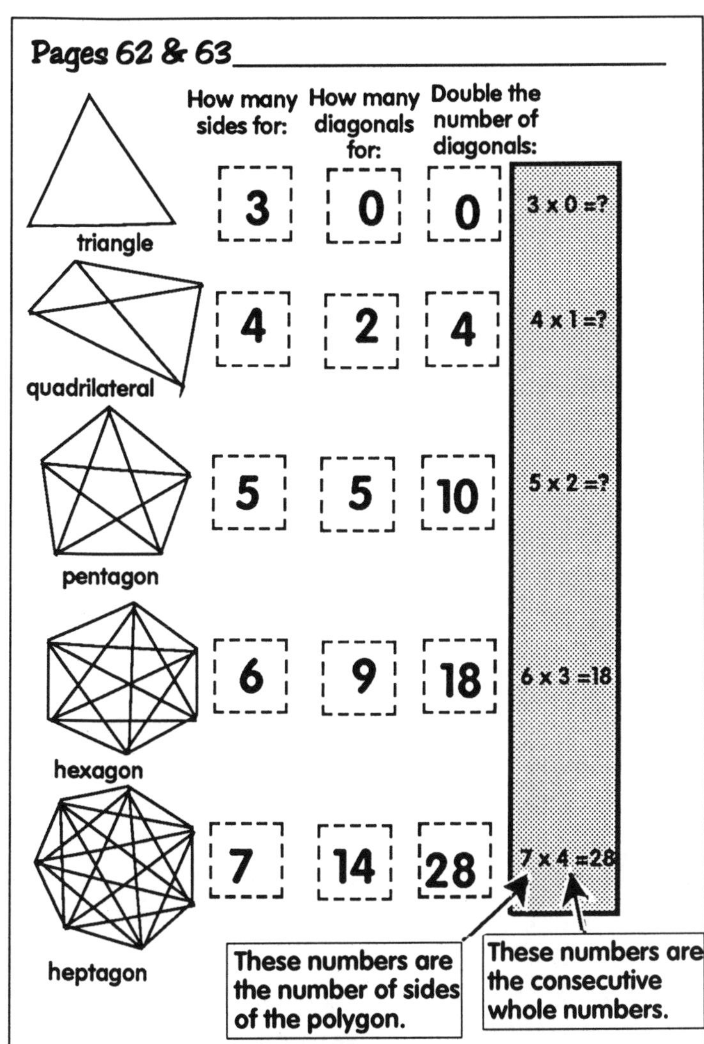

Notice that the product of these two numbers always equals twice the number of diagonals for a polygon. Taking half of this product will give the number of diagonals for any particular polygon. Now notice that the consecutive whole number that goes along with the number of sides is always 3 less than the number of sides. For example, a heptagon has 7 sides, the consecutive whole number to multiply 7 by is found by 7-3=4, so the number of diagonals for a heptagon would be half of 7x4 or half of 28, namely 14!

Page 50

1) These number form the falling equation:

$$9 + 16 = 25$$

2) $25 + 144 = 169$

3) No, sides with lengths 4, 5, and 6 do not make a right triangle. No their squares do not form an equation,

$$16 + 25 \neq 36.$$

Page 55

Euler noticed each landing point had either an odd or even number of bridges to it. He reasoned if a diagram had more than two odd landing points it could not be traced with a pencil without doubling back. *Why?* Because an odd point was formed whenever a path began or ended there. Since there can be at most one beginning and one ending point, there cannot be more than two odd landing points. The Königsberg bridge problem has 3 odd landing points— the upper bank of town has 3 bridges to it, the Kneiphof island has 5 bridges to it , and the lower bank of town has 3 bridges to it.

Page 68

twenty-four nonillion, seven-hundred octillion, six-hundred and thirty-one septillion, five-hundred sextillion, seventy quintillion, two quadrillion, one-hundred trillion, three-hundred and forty-four billion, two-hundred and ninety-eight million, twenty-three thousand, one-hundred and eighty-five.

answers & solutions

Page 73

Put the button hole through the loop until the pencil point can pass through the button hole. Pull the pencil through the button hole.

Reverse the steps above to remove the pencil from the button hole.

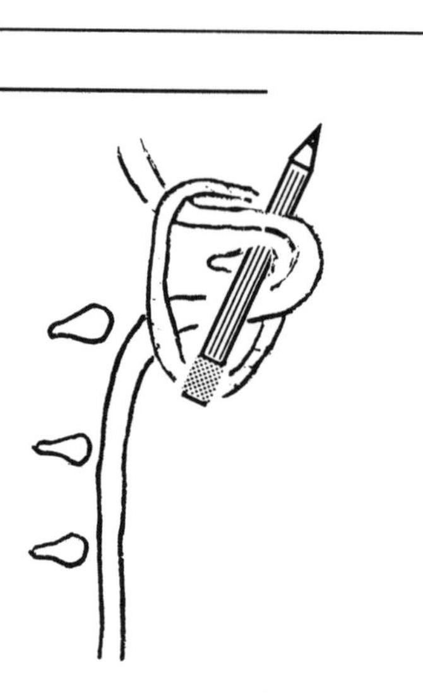

Pages 76 & 77

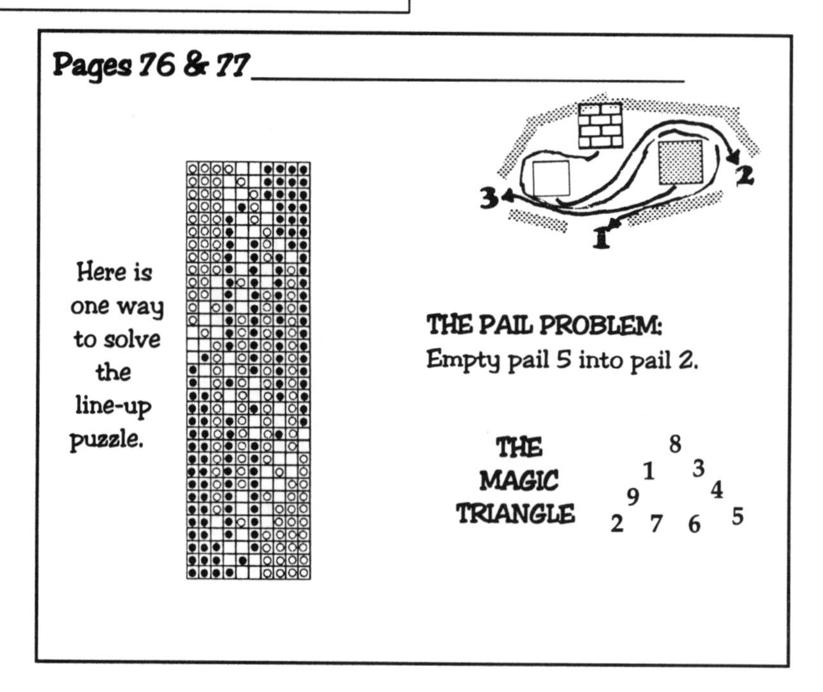

Here is one way to solve the line-up puzzle.

THE PAIL PROBLEM:
Empty pail 5 into pail 2.

THE MAGIC TRIANGLE

8
1 3
9 4
2 7 6 5

answers & solutions

Pages 82 & 83

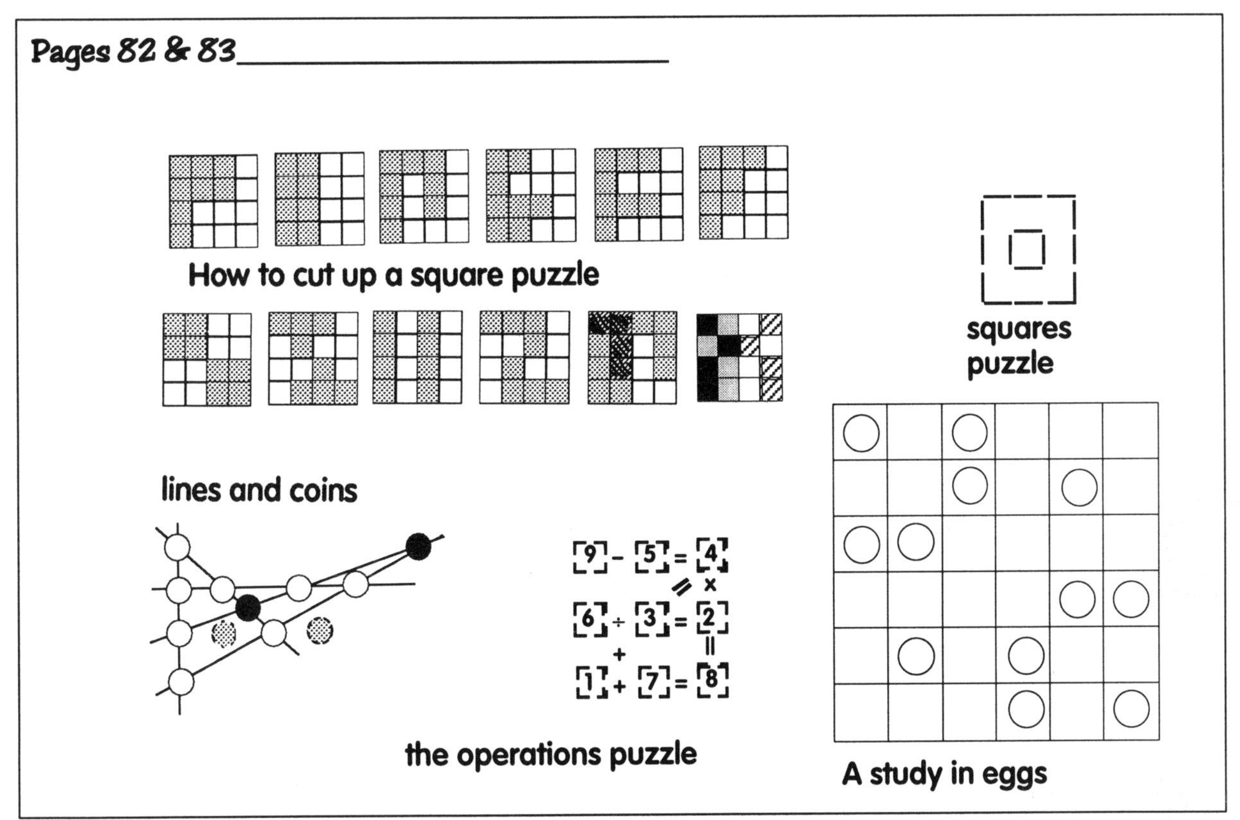

Pages 86 & 87

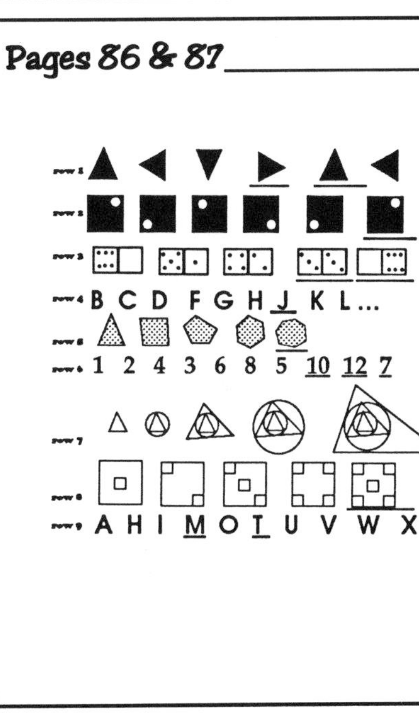

(row 1) The triangle is turning counterclockwise.

(row2) The dot's movement always follows this pattern: its first move is always to the opposite corner and its second move is always to the clockwise corner.

(row 3) The dots on the dominoes follow the number patterns: 6+0; 5+1; 4+2; 3+3; 2+4; 1+5; 0+6.

(row 4) Only the consonants are listed in alphabetical order.

(row 5) Each new polygon has one more side than the previous one.

(row 6) The pattern is: the first odd number is followed by the first two even numbers, then the next odd number is followed by the next two even numbers, and so forth.

(row 7) The pattern is: a triangle , then a circle is placed around it, then a triangle is placed around that circle, then a circle is placed around that triangle and so forth.

(row 8) The little squares added follow the number pattern: 1,2,3,4,5,6,...

(row 9) The letters which do not change when reflected in the mirror are listed in alphabetical order.

answers & solutions

Pages 88 & 89

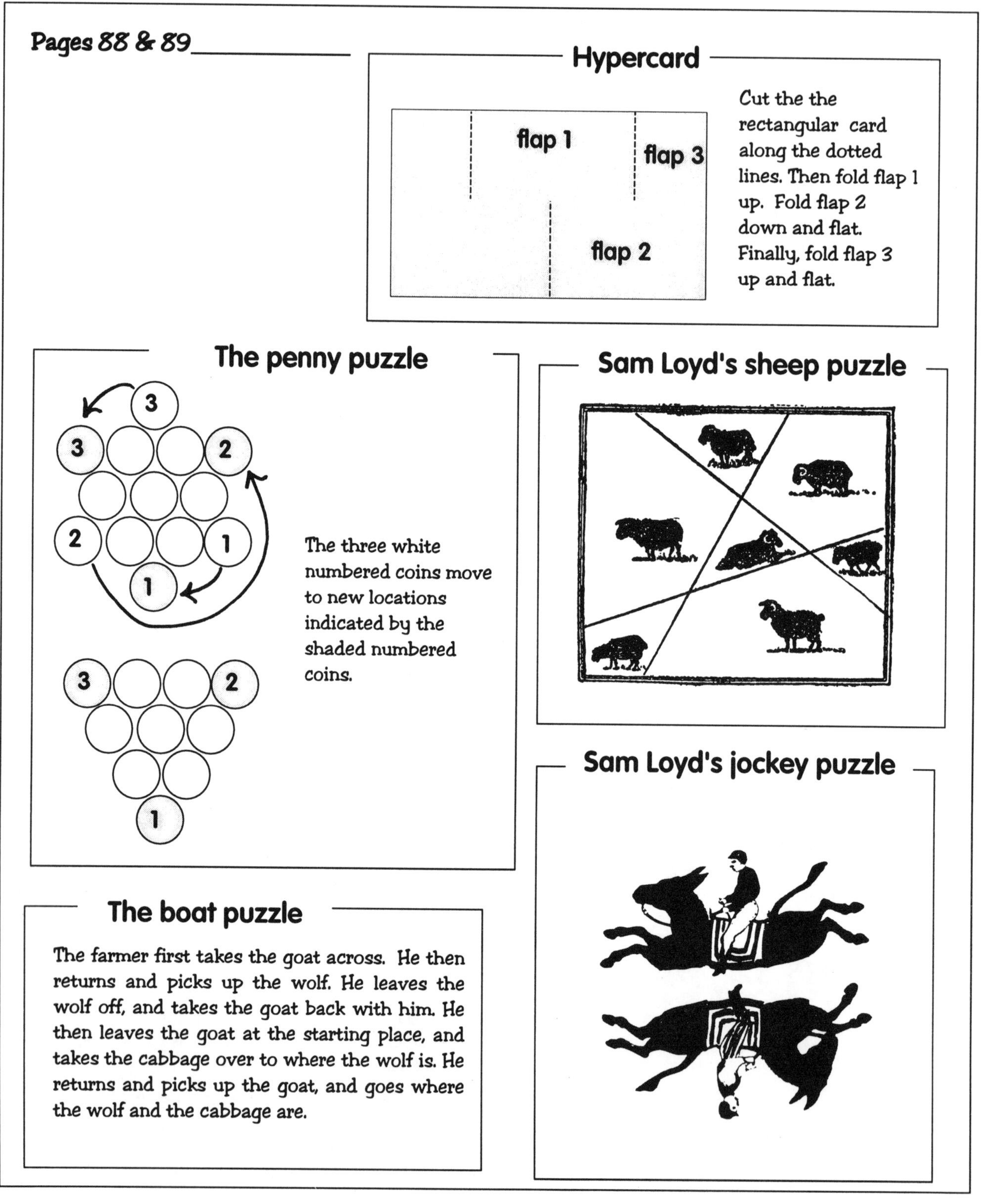

The boat puzzle

The farmer first takes the goat across. He then returns and picks up the wolf. He leaves the wolf off, and takes the goat back with him. He then leaves the goat at the starting place, and takes the cabbage over to where the wolf is. He returns and picks up the goat, and goes where the wolf and the cabbage are.

answers & solutions

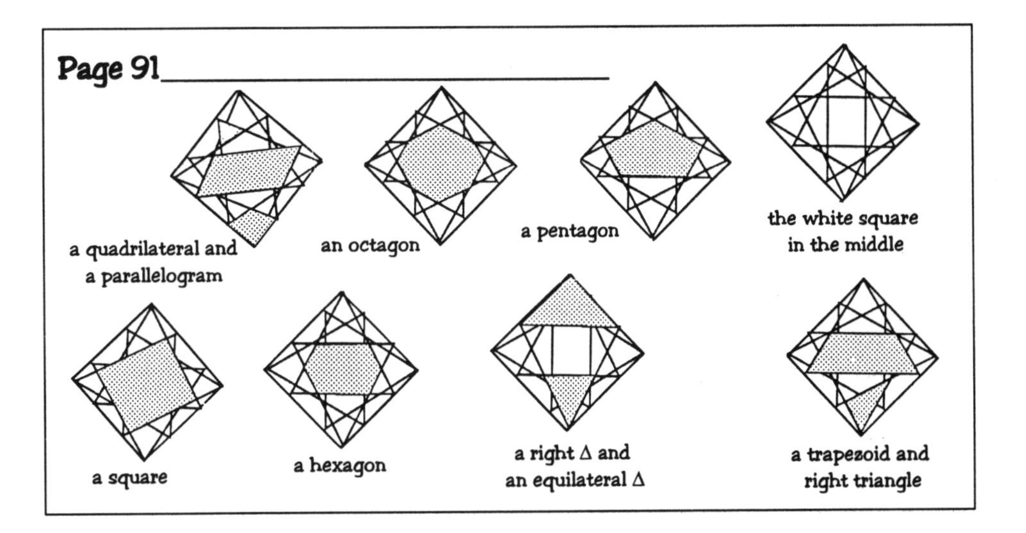

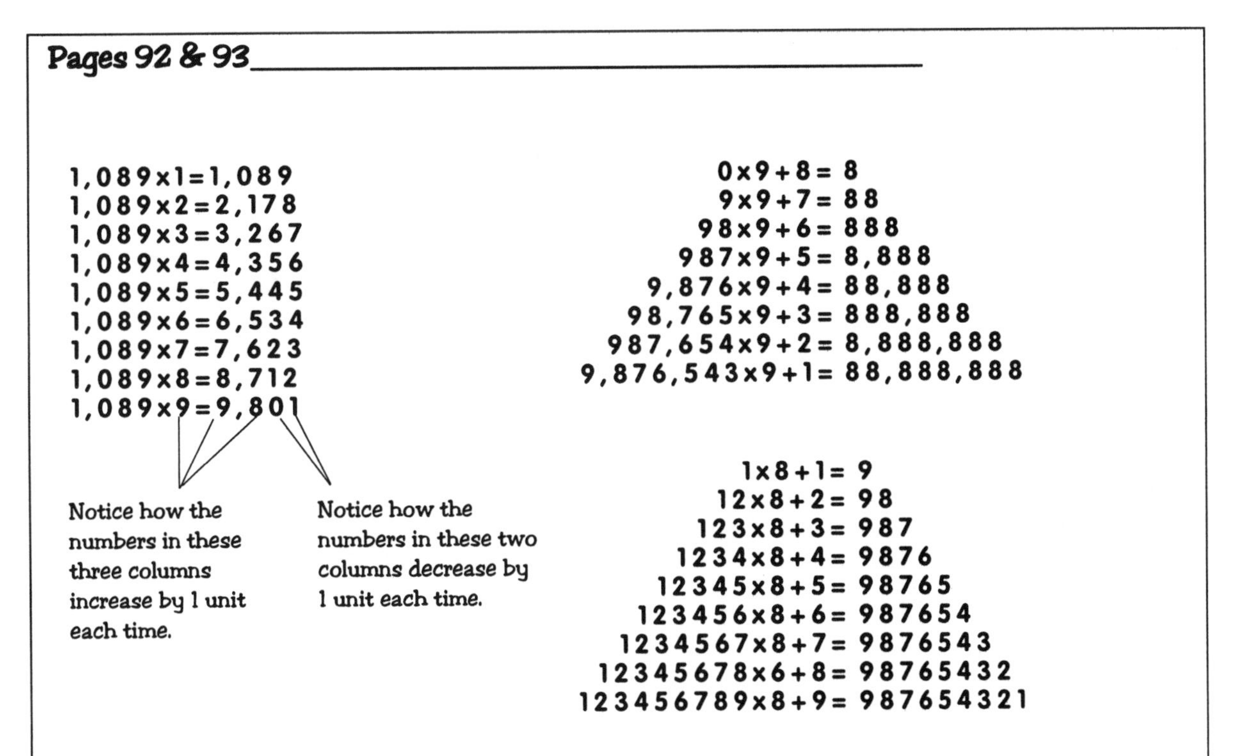

Pages 92 & 93

1,089x1=1,089
1,089x2=2,178
1,089x3=3,267
1,089x4=4,356
1,089x5=5,445
1,089x6=6,534
1,089x7=7,623
1,089x8=8,712
1,089x9=9,801

Notice how the numbers in these three columns increase by 1 unit each time.

Notice how the numbers in these two columns decrease by 1 unit each time.

0x9+8= 8
9x9+7= 88
98x9+6= 888
987x9+5= 8,888
9,876x9+4= 88,888
98,765x9+3= 888,888
987,654x9+2= 8,888,888
9,876,543x9+1= 88,888,888

1x8+1= 9
12x8+2= 98
123x8+3= 987
1234x8+4= 9876
12345x8+5= 98765
123456x8+6= 987654
1234567x8+7= 9876543
12345678x6+8= 98765432
123456789x8+9= 987654321

answers & solutions

Pages 94 & 95

Thes hat problem:

ANSWER: If Jerry had on a black hat, then Tom would have known that he had a tan one, because there is only 1 black hat. Since Tom could not answer the question, Jerry knew his hat had to be tan.

The small change problem:

ANSWER:

1-half dollar, 1-quarter and 3-dimes.

- You cannot have 2 half dollars, because you cannot make change for a dollar.
- You cannot have two quarters because you cannot make change for half dollar.
- You cannot have two nickels because you cannot make change for a dime.
- You cannot have five pennies because you cannot make change for a nickel.
- You can have up to 4 dimes.

How did the number get there?

EXPLANATION: Look at the number 358. The number 3 is in the hundreds place, the 5 is in the tens place and the 8 is in the ones place. The steps outlined, end up multiplying the 1st digit by 100, the 2nd by 10 and the 3rd by 1, which places the digits in a number in the order the digits were chosen.

Arranging the line-up

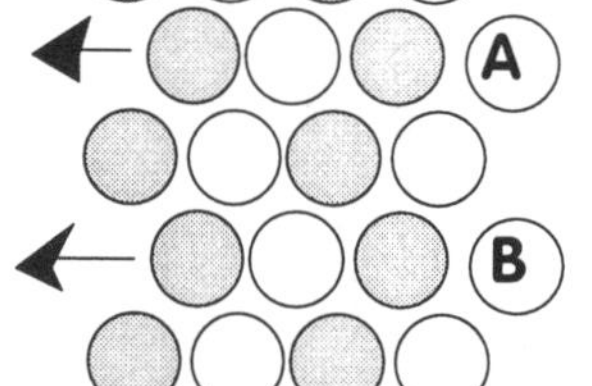

Use checkers A and B to push over the two rows as shown.

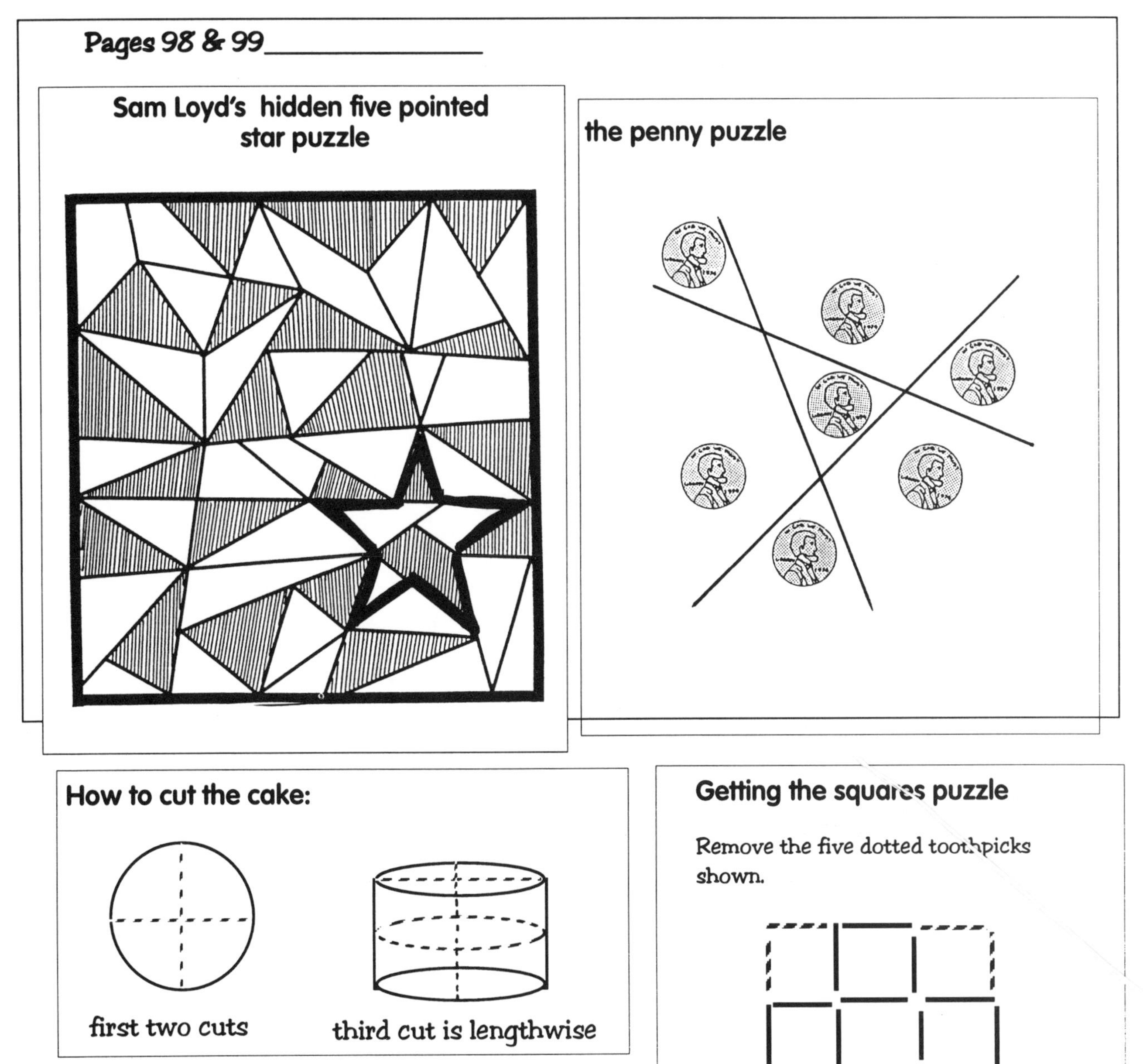
Pages 98 & 99
Sam Loyd's hidden five pointed star puzzle
the penny puzzle
How to cut the cake:
first two cuts
third cut is lengthwise
Getting the squares puzzle
Remove the five dotted toothpicks shown.

answers & solutions

Pages 100 & 101

Where's the missing dollar puzzle.
There is no missing dollar. One just needs to keep track of the amounts paid, and where they are located. $10 dollars is in the register. Each of the three friends got $1 and the clerk got $2. That totals $15.

Where's the missing man puzzle.
Try to see which warrior is lost. As you see, they share parts of their bodies. Study the Earth carefully as it is rotated, and you'll see how one warrior becomes part of an existing one.

Where's the missing desk puzzle.
When the librarian began to seat the six students, she started with the 2nd student rather than the 1st. She was probably considering Tom the 1st student, but Tom is the 7th student.

answers & solutions

Pages 102 & 103

1) Only one. They were only born once.

2) The 4th of July is a date which happens everywhere each year throughout the world.

3) A man can't live in San Francisco and be buried in New York City because he is still living.

4) Any size dog can only run half way into the woods because after it has passed the half way mark it is running out of the woods.

5) Each won two and they tied one game.

6) I would have to light the match first so I could light the other things.

7) I would have swallowed the last pill 60 minutes (or an hour) after the first pill.

8) She knew it was counterfiet because it had B.C. (Before Christ) marked on it. How could it be marked B.C., if Christ had not been born yet?

9) Each inning has six outs. Three for each team.

10) The nurse is Mary's brother.

11) They stand back to back.

12) They weigh the same, one pound each.

13) It would be impossible for a man in Oregon to marry his widow's sister because he would have to be dead if his wife were a widow.

14) Eight days.

15) One pile of sand.

16) You have two apples in your hands.

17) The hole has no earth in it. It is a hole.

18) The other is a penny and the one in your hand is a quarter.

19) All months of the year will have 29 days.

Pages 104 & 105

When model A is cut along the dotted lines, it results in a square.

When model B is cut along the dotted lines, it falls apart into two separate figures.

children's mathematics books by theoni pappas

FRACTALS, GOOGOLS and Other Mathematical tales

A treasure trove of Penrose's adventures and other stories that make mathematical ideas come to life. Explore math concepts with Penrose the cat, and other mathematical characters—such as π, the numberline, Leonhard the magic turtle, the googols, Fibonacci rabbit. Offers an amusing and entertaining way to explore and share mathematical ideas regardless of age or math background.

$9.95 • 64 pages • ISBN:0-933174-89-6

MATH FOR KIDS & OTHER PEOPLE TOO!

by Theoni Pappas

Join Penrose on a mad cap tour of mathematical ideas.

- Venture with him when he discovers how to help the ÷2
- Meet the fractal dragon
- Watch a tangram egg hatch
- learn how to make a square become a bird
- help nanocat get back home
- and many more amusing, entertaining and informative tales.

All told in an enchanting and captivating style that is sure to make mathematics fun as well as educational. The reader will gain new insights and appreciation for mathematics and its many facets.

$10.95 • 128 pages • ISBN:1-884550-14-2

THE ADVENTURES OF PENROSE The Mathematical Cat

by Theoni Pappas

Join Penrose on a madcap tour of mathematical ideas.

- Venture with him when he discovers how to help the square root of 2
- Meet the fractal dragon
- Watch a tangram egg hatch
- learn how to make a square become a bird
- Help nanocat get back home
- and many more amusing, entertaining and informative tales.

All told in an enchanting and captivating style that is sure to make mathematics fun as well as educational. The reader will ain new insights and appreciation for mathematics and its any facets.

$10.95 • 128 pages • ISBN:1-884550-14-2

other mathematics books by theoni pappas

THE JOY OF MATHEMATICS & MORE JOY OF MATHEMATICS

by Theoni Pappas

Part of the joy of mathematics is that it is everywhere—in soap bubbles, electricity, da Vinci's masterpieces, even in an ocean wave. The book has 200 clearly illustrated mathematical ideas, concepts, puzzles, games that show where they turn up in the "real" world. You'll find out what a googol is, visit hotel infinity, read a thorny logic problem that was stumping them back in the 8th century. From the paradox of the Unexpected Exam to Lewis Carroll's decidedly Tangled Tale, THE JOY OF MATHEMATICS adds up to the most fun you've had in a long time.

THE JOY OF MATHEMATICS
•NEWLY REVISED EDITION
•256 pages•illustrations•b/w photos •diagrams•5 1/2"x8 1/2"
•ISBN:0-933174-65-9 • **$10.95**

"... designed to help the reader become aware of the inseparable relationship of mathematics and the world by presenting glimpses and images of mathematics in the many facets of our lives."

—Science News

"...helps demystify this often misunderstood subject...randomly arranged topics retain the true essence of discovery...test your wits with topo,a mathematical game; ponder the Trinity of Rings; or juggle with a frustrating puzzle—this book gives glimpses of the many places math exists, helping you discover some of its treasures."

—Curriculum Product Review

Featured selection of
QUALITY PAPERBACK BOOK CLUB®
BOOK-OF-THE-MONTH CLUB™
Pick Of The Paperbacks

— Over 200,000 copies in print.—

MORE JOY OF MATHEMATICS *is a sequel with another treasure trove of all new topics in the same inviting format as THE JOY OF MATHEMATICS.*

"Approach goes far beyond the usual mechanical presentation, linking math to vital, everyday life and adding cultural and historical influences to connect math concepts to realistic challenges...fun rather than tedious."

—The Bookwatch

MORE JOY OF MATHEMATICS
•296 pages•illustrations•b/w photos •diagrams
•5 1/2"x8 1/2"
•ISBN:0-933174-73-X • **$10.95**

Featured selection of
QUALITY PAPERBACK BOOK CLUB®
BOOK-OF-THE-MONTH CLUB™

other mathematics books by theoni pappas

$10.95 •8.5"x5.5"• photos & illustrations throughout • 324 pages
• ISBN:0-933174-99-3

Selected by
QUALITY PAPERBACK BOOK CLUB®
Book-of-the-MONTH CLUB

THE MAGIC OF MATHEMATICS

by Theoni Pappas

Delves into the world of ideas, explores the spell mathematics casts on our lives, and helps you discover mathematics where you least expect it.

"Thought-provoking yet not overwhelming, this book makes the author's pleasure in mathematics evident." **—Library Journal**

"...witty and intriguing..." **—Science News**

"Readers will relish this book. The focus is to explain math in a lively, revealing manner. It succeeds!" **—The Bookwatch**

"Pappas writes at a level that puts all her topics within reach of any mathematics teacher and usually within reach of a generalist as well. In spite of this, she does not skim over facts or misrepresent them. If you haven't bought Pappas' first two books, then get them. If you have them, buy this one as well.

—The Mathematics Teacher

THE MUSIC OF REASON

THE BEAUTY OF MATHEMATICS THROUGH QUOTATIONS

by Theoni Pappas

$9.95
• 138 pp
• ISBN: 1-884550-045

Learn what Alice in Wonderland, Albert Einstein, William Shakespeare, Mae West, Plato and others have to say about mathematics.

THE MUSIC OF REASON is a compendium of profound and profane thoughts on mathematics by mathematicians, scientists, authors and artists.

"The collection of quotes is divided into fifteen categories...each preceded by a short commentary...(which) act as a springboard for some lively discussions linking mathematics to other subjects and to the real world. This amusing little book will complement the collection of mathematicians and mathematics educators alike.

—The Mathematics Teacher.

Theoni Pappas

About the Author

Mathematics teacher and consultant Theoni Pappas received her B.A. from the University of California at Berkeley in 1966 and her M.A. from Stanford University in 1967. Pappas is committed to demystifying mathematics and to helping eliminate the elitism and fear often associated with it.

In addition to *Math for Kids & Other People Too!,* her other innovative creations include *The Children's Mathematics Calendar, The Math-T-Shirt, The Mathematics Calendar, The Mathematics Engagement Calendar,* and *What Do You See?* —An optical illusion slide show with text. Pappas is also the author of the following books: *Mathematics Appreciation, The Joy of Mathematics, Greek Cooking for Everyone, Math Talk, More Joy of Mathematics, Fractals, Googols & Other Mathematical Tales, The Magic of Mathematics, The Music of Reason, Mathematical Scandals, The Adventures of Penrose —The Mathematical Cat, Math-A-Day* and *Mathematical Footprints.*